Structural Mechanics with a Pen

Andreas Öchsner

Structural Mechanics with a Pen

A Guide to Solve Finite Difference Problems

 Springer

Andreas Öchsner
Faculty of Mechanical Engineering
Esslingen University Applied Sciences
Esslingen am Neckar, Baden-Württemberg
Germany

ISBN 978-3-030-65894-6   ISBN 978-3-030-65892-2   (eBook)
https://doi.org/10.1007/978-3-030-65892-2

This Springer imprint is published by the registered company Springer Nature Switzerland AG
The registered company address is: Gewerbestrasse 11, 6330 Cham, Switzerland

Science knows no country, because knowledge belongs to humanity, and is the torch which illuminates the world.

Louis Pasteur (1822–1895)

Preface

The *classical* approximation methods known in continuum mechanics are the finite difference method, the finite element method, the finite volume method, and the boundary element method. Each method has its advantages and disadvantages and a different spread in different areas of applied mechanics, e.g. solid or fluid mechanics. The oldest approximation method to solve partial differential equations is the finite difference method since the mathematical set of tools is basically limited to the series expansion of derivatives. More precisely, truncated Taylor's series is used for the local expansions of the variables. This 'simple' derivation of the method makes it quite attractive to introduce approximation methods in tertiary engineering education, e.g. in mechanical or civil engineering. This allows acquiring a general understanding of numerical approximation methods and the involved common steps such as discretization, assembly of a global system of equations, consideration of boundary and load conditions as well as the solution of a linear or nonlinear system of equations.

This book is focused on the introduction of the finite difference method based on the classical one-dimensional structural members, i.e. rods/bars and beams. It is the goal to provide a first introduction to the manifold aspects of the finite difference method and to enable the reader to get a methodical understanding of important subject areas in structural mechanics. The reader learns to understand the assumptions and derivations of different structural members. Furthermore, she/he learns to critically evaluate the possibilities and limitations of the finite difference method. Additional comprehensive mathematical descriptions, which solely result from advanced illustrations for two- or three-dimensional problems, are omitted. Hence, the mathematical description largely remains simple and clear.

Chapter 1 illustrates the derivation of the finite difference method in a general way for one-dimensional problems, i.e. partial differential equations. Chapter 2 covers the simplest one-dimensional element type, i.e. the rod/bar element. Approximate equations are provided for rods of constant and varying tensile stiffnesses. Chapter 3 covers the simplest one-dimensional beam formulation according to Euler-Bernoulli. This element is also called the thin beam. Again, approximate equations are provided for thin beams of constant and varying bending stiffnesses. Chapter 4 treats a higher beam bending theory according to Timoshenko. This theory considers the contribution of the shear force on the deformation. Chapter 5 introduces a simple treatment of elasto-plastic bending problems under the consideration of the thin beam formulation. The so-called layered approach is applied to linear-elastic/ideal-plastic material behavior and is restricted to monotonic loading.

All derivations in chapters two to four follow a common approach: based on the three basic equations of continuum mechanics, i.e. the kinematics relationship, the constitutive law, and the equilibrium equation, the partial differential equations, which describe the physical problem, are presented. The finite difference method is then used to derive approximate equations for the corresponding structural member.

In order to deepen the understanding of the derived equations and theories, each technical chapter collects at its end supplementary calculation problems. A short solution for each problem is included at the end of this book. It should be noted that these short solutions contain major steps for the solution of the problem and not only, for example, a numerical value for the final result. This should ensure that students are able to successfully master these problems. I hope that students find this book a useful complement to many classical textbooks. I look forward to receiving their comments and suggestions.

Esslingen, Germany   Andreas Öchsner
October 2020

Acknowledgements

It is important to highlight the contribution of the students which helped to develop the concept and content of this book. Their questions, comments, and struggles during difficult lectures, assignments, and final exams helped me to develop the idea for this approach. In addition, it is important to highlight the contribution of many undergraduate and postgraduate students. Furthermore, I would like to express my sincere appreciation to the Springer publisher, especially to Dr. Christoph Baumann, for giving me the opportunity to realize this book.

Contents

Symbols and Abbreviations

Latin Symbols (Capital Letters)

A	Area, cross-sectional area
A_s	Shear area
C	Elasticity constant
D	Bending rigidity
$\boldsymbol{D}$	Compliance matrix
E	Young's modulus
EA	Tensile stiffness
EI	Bending stiffness
F	Force
$\boldsymbol{F}$	Column matrix of loads
G	Shear modulus
I	Second moment of area
$\boldsymbol{K}$	Global stiffness matrix
$\boldsymbol{K}_T$	Tangent stiffness matrix
L	Length
M	Moment
N	Normal force (internal), Interpolation function
Q	Shear force (internal)
R	Equivalent nodal force
W	Weight function
X	Global Cartesian coordinate
ΔX	Node spacing
Y	Global Cartesian coordinate
Z	Global Cartesian coordinate

Latin Symbols (Small Letters)

a Geometric ratio, geometric dimension
b Body force, geometric dimension
$\boldsymbol{b}$ Column matrix of body forces acting per unit volume
c Constant of integration
e Column matrix of generalized strains
f Scalar function
h Geometric dimension
i Actual node number, iteration index
j Iteration index
k Elastic embedding modulus, auxiliary function, elastic foundation modulus, yield stress, layer number
k_s Shear correction factor
k_t Tensile yield stress
m Distributed moment
n Total number of nodes, Iteration number
p Distributed load in X-direction
q Distributed load in Z-direction
r Residual
$\boldsymbol{r}$ Residual column matrix
s Column matrix of generalized stresses
t Time
u Displacement
$\boldsymbol{u}$ Column matrix of displacements, Column matrix of nodal unknowns
v Auxiliary function

Greek Symbols (Small Letters)

α Parameter
β Parameter
γ Shear strain (engineering definition)
ϵ Strain
κ Curvature
ν Poisson's ratio
ξ Natural coordinate
σ Stress
τ Shear stress
ϕ Rotation (Timoshenko beam)
φ Rotation (Bernoulli beam)

Mathematical Symbols

$\times$	Multiplication sign (used where essential)
$\mathcal{L}\{\dots\}$	Differential operator
$\mathcal{L}$	Matrix of differential operators
$O(\dots)$	Order of
δ	Dirac delta function

Indices, Superscripted

$\dots^{\text{init}}$	Initial
$\dots^{\text{el}}$	Elastic
$\dots^{\text{elpl}}$	Elasto-plastic
$\dots^{\text{pl}}$	Plastic

Indices, Subscripted

$\dots_{\text{c}}$	Compression, center
$\dots_{\text{max}}$	Maximum
$\dots_{\text{t}}$	Tensile

Abbreviations

1D	One-dimensional
BC	Boundary condition
BD	Backward difference
CD	Centered difference
FD	Forward difference, Finite difference
FDM	Finite difference method
PDE	Partial differential equation

Chapter 1
Idea and Derivation of the Method

The first widely known approximation method for partial differential equations is the finite difference method [1, 11] which approximates the governing differential equations of a field problem using local expansions for the variables, generally truncated Taylor's series. Comprehensive descriptions of the method can be found, for example, in [4, 5, 7]. Other classical approximation methods are the finite element method [9, 10], the finite volume method, and the boundary element method.

As a first step to solve a differential equation for a one-dimensional problem, it is assumed that the numerical solution $u(X)$ is to be determined only at a set of n points within the domain $X \in [0, L]$ including the ends, cf. Fig. 1.1.

These n nodes or grid points will be used to derive approximations to the derivatives of the function $u(X)$. To simplify the approach, let us assume in the following that these n nodes are equally spaced, at a distance equal to[1] $\Delta X = \frac{L}{n-1}$. As in the case of the finite element method, high gradients would require a smaller node spacing in order to meet the requirements on the accuracy of the approximate solution. However, the general idea of the finite difference method will be here introduced based on an equi-spaced division of the domain $0 \leq X \leq L$. A typical inner node of the domain is denoted in the following by i and the two neighboring nodes are called $i - 1$ on the left-hand and $i + 1$ on the right-hand side. For sufficient smooth functions $u(X)$, a Taylor's series expansion around node i gives:

$$u_{i+1} = u_i + \left(\frac{\mathrm{d}u}{\mathrm{d}X}\right)_i \Delta X + \left(\frac{\mathrm{d}^2 u}{\mathrm{d}X^2}\right)_i \frac{\Delta X^2}{2} + \cdots + \left(\frac{\mathrm{d}^k u}{\mathrm{d}X^k}\right)_i \frac{\Delta X^k}{k!}, \tag{1.1}$$

$$u_{i-1} = u_i - \left(\frac{\mathrm{d}u}{\mathrm{d}X}\right)_i \Delta X + \left(\frac{\mathrm{d}^2 u}{\mathrm{d}X^2}\right)_i \frac{\Delta X^2}{2} - \cdots + \left(\frac{\mathrm{d}^k u}{\mathrm{d}X^k}\right)_i \frac{\Delta X^k}{k!}. \tag{1.2}$$

---

[1]It should be noted here that the first node in this derivation is denoted by '1' and not as in some references as '0'.

© The Author(s), under exclusive license to Springer Nature Switzerland AG 2021
A. Öchsner, *Structural Mechanics with a Pen*,
https://doi.org/10.1007/978-3-030-65892-2_1

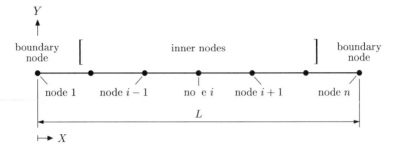

Fig. 1.1 Finite difference model of a one-dimensional problem

The infinite series of Eqs. (1.1) and (1.2) are truncated for practical use after a certain number of terms. As a result of this approximation, the so-called truncation errors occurs. Summing up the expression of Eqs. (1.1) and (1.2) gives

$$u_{i+1} + u_{i-1} = 2u_i + \left(\frac{d^2u}{dX^2}\right)_i \Delta X^2 + \frac{1}{12}\left(\frac{d^4u}{dX^4}\right)_i \Delta X^4 + \cdots , \tag{1.3}$$

or rearranged for the second order derivative:

$$\left(\frac{d^2u}{dX^2}\right)_i = \frac{u_{i+1} - 2u_i + u_{i-1}}{\Delta X^2} \underbrace{- \frac{1}{12}\left(\frac{d^4u}{dX^4}\right)_i \Delta X^2 - \cdots}_{O(\Delta X^2)} . \tag{1.4}$$

The symbol 'O' in Eq. (1.4) reads 'order of' and states that if the second order derivative of $u(X)$ is approximated by the first expression on the right-hand side of Eq. (1.4), then the truncation error is of order of ΔX^2. This approximation is a second order accurate approximation because of the truncated terms and is called the centered difference scheme. In a similar way, other derivatives can be derived from Eqs. (1.1) and (1.2). Subtracting of Eq. (1.2) from (1.1) gives

$$u_{i+1} - u_{i-1} = 2\left(\frac{du}{dX}\right)_i \Delta X + \frac{1}{3}\left(\frac{d^3u}{dX^3}\right)_i \Delta X^3 + \cdots \tag{1.5}$$

or rearranged for the first order derivative

$$\left(\frac{du}{dX}\right)_i = \frac{u_{i+1} - u_{i-1}}{2\Delta X} \underbrace{- \frac{1}{6}\left(\frac{d^3u}{dX^3}\right)_i \Delta X^2 - \cdots}_{O(\Delta X^2)} . \tag{1.6}$$

This last approximation of the first order derivative is called centered difference or centered Euler and the truncation error is of order ΔX^2. Rearranging Eq. (1.1), i.e.

$$u_{i+1} - u_i = \left(\frac{du}{dX}\right)_i \Delta X + \left(\frac{d^2u}{dX^2}\right)_i \frac{\Delta X^2}{2} + \cdots, \tag{1.7}$$

or

$$\left(\frac{du}{dX}\right)_i = \frac{u_{i+1} - u_i}{\Delta X} - \underbrace{\left(\frac{d^2u}{dX^2}\right)_i \Delta X - \cdots}_{O(\Delta X)}, \tag{1.8}$$

which gives an expression for the forward difference or forward EULER approximation of the first order derivative. In a similar way, Eq. (1.2) can be rearranged to obtain the backward difference or backward EULER approximation of the first order derivative as:

$$\left(\frac{du}{dX}\right)_i = \frac{u_i - u_{i-1}}{\Delta X} + \underbrace{\left(\frac{d^2u}{dX^2}\right)_i \Delta X - \cdots}_{O(\Delta X)}. \tag{1.9}$$

Graphical representations of the finite difference approximations of first order derivatives are shown in Fig. 1.2.

Further derivatives of $u(X)$ of different order can be derived if u is evaluated at nodes $i + 2, i - 2$ etc. through a TAYLOR's series about the node i. Finite difference formulae of the first order derivative with truncation error of order ΔX^2, based on a forward or backward difference approximation, can be obtained in the following way: Let us consider in Eq. (1.2) the second order derivative, i.e.

$$u_{i-1} - u_i = -\left(\frac{du}{dX}\right)_i \Delta X + \frac{1}{2}\left(\frac{d^2u}{dX^2}\right)_i \Delta X^2 - \frac{1}{6}\left(\frac{d^3u}{dX^3}\right)_i \Delta X^3 + \cdots \tag{1.10}$$

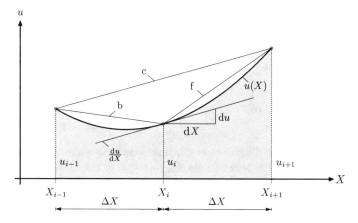

Fig. 1.2 Graphical representation of first order derivative approximations: c: centered; f: forward; b: backward difference

or

$$\left(\frac{du}{dX}\right)_i = \frac{u_i - u_{i-1}}{\Delta X} + \frac{1}{2}\left(\frac{d^2 u}{dX^2}\right)_i \Delta X \underbrace{- \frac{1}{6}\left(\frac{d^3 u}{dX^3}\right)_i \Delta X^2 + \cdots,}_{O(\Delta X^2)} \tag{1.11}$$

where the second order derivative can be replaced by a backward approximation[2] with truncation error of order ΔX^2, i.e.

$$\left(\frac{d^2 u}{dX^2}\right)_i = \frac{u_i - 2u_{i-1} + u_{i-2}}{\Delta X^2} + \left(\frac{d^3 u}{dX^3}\right)_i \Delta X - O(\Delta X^2). \tag{1.12}$$

Thus, the backward approximation of the first order derivative with truncation error of order ΔX^2 is obtained as:

$$\left(\frac{du}{dX}\right)_i = \frac{u_i - u_{i-1}}{\Delta X} + \frac{u_i - 2u_{i-1} + u_{i-2}}{2\Delta X} + \frac{1}{2}\left(\left(\frac{d^3 u}{dX^3}\right)_i \Delta X^2 - O(\Delta X^3)\right)$$

$$= \frac{3u_i - 4u_{i-1} + u_{i-2}}{2\Delta X} + O(\Delta X^2). \tag{1.13}$$

Some common expressions for derivatives of different order and the respective truncation errors are summarized in Table 1.1.

It is possible to derive the finite difference approximations based on a special type of collocation method [3]. This so-called cell collocation method [6] considers a cell as shown in Fig. 1.3. Let us remind here that the point collocation method uses as weight function the Dirac delta function, i.e.,

$$\delta(X - X_k) = \begin{cases} 0 & \text{for } X \neq X_k \\ \infty & \text{for } X = X_k \end{cases}. \tag{1.14}$$

To approximate, for example, a second order derivative $\frac{d^2 u}{dX^2}$, a local approximate function can be written as

$$u = \sum_{k=i-1}^{i} N_k \times u_k = N_{i-1} u_{i-1} + N_i u_i + N_{i+1} u_{i+1}, \tag{1.15}$$

where $u_{i-1}, \ldots, u_{i+1}$ are the values of the function at the nodes. The interpolation functions N_k are given in natural coordinates $(-1 \leq \xi \leq 1)$ and can take the following quadratic form (cf. Fig. 1.4), [3]:

---

[2]This formulation can be obtained based on a TAYLOR's series expansion for $i - 2$ up to $\frac{d^6 u}{dX^6}$ and introducing a backward difference approximation of the first order derivative which contains the terms up to $\frac{d^6 u}{dX^6}$.

Table 1.1 Finite difference approximations for various differentiations, partly taken from [2, 4]

Derivative	Finite difference approximation	Type	Error
$\left(\dfrac{du}{dX}\right)_i$	$\dfrac{u_{i+1}-u_i}{\Delta X}$	Forward diff.	$O(\Delta X)$
	$\dfrac{-3u_i+4u_{i+1}-u_{i+2}}{2\Delta X}$	"	$O(\Delta X^2)$
	$\dfrac{u_i-u_{i-1}}{\Delta X}$	Backward diff.	$O(\Delta X)$
	$\dfrac{3u_i-4u_{i-1}+u_{i-2}}{2\Delta X}$	"	$O(\Delta X^2)$
	$\dfrac{u_{i+1}-u_{i-1}}{2\Delta X}$	Centered diff.	$O(\Delta X^2)$
$\left(\dfrac{d^2u}{dX^2}\right)_i$	$\dfrac{u_{i+2}-2u_{i+1}+u_i}{\Delta X^2}$	Forward diff.	$O(\Delta X)$
	$\dfrac{-u_{i+3}+4u_{i+2}-5u_{i+1}+2u_i}{\Delta X^2}$	"	$O(\Delta X^2)$
	$\dfrac{u_i-2u_{i-1}+u_{i-2}}{\Delta X^2}$	Backward diff.	$O(\Delta X)$
	$\dfrac{2u_i-5u_{i-1}+4u_{i-2}-u_{i-3}}{\Delta X^2}$	"	$O(\Delta X^2)$
	$\dfrac{u_{i+1}-2u_i+u_{i-1}}{\Delta X^2}$	Centered diff.	$O(\Delta X^2)$
$\left(\dfrac{d^3u}{dX^3}\right)_i$	$\dfrac{u_{i+3}-3u_{i+2}+3u_{i+1}-u_i}{\Delta X^3}$	Forward diff.	$O(\Delta X)$
	$\dfrac{-3u_{i+4}+14u_{i+3}-24u_{i+2}+18u_{i+1}-5u_i}{2\Delta X^3}$	"	$O(\Delta X^2)$
	$\dfrac{u_i-3u_{i-1}+3u_{i-2}-u_{i-3}}{\Delta X^3}$	Backward diff.	$O(\Delta X)$
	$\dfrac{5u_i-18u_{i-1}+24u_{i-2}-14u_{i-3}+3u_{i-4}}{2\Delta X^3}$	"	$O(\Delta X^2)$
	$\dfrac{u_{i+2}-2u_{i+1}+2u_{i-1}-u_{i-2}}{2\Delta X^3}$	Centered diff.	$O(\Delta X^2)$
$\left(\dfrac{d^4u}{dX^4}\right)_i$	$\dfrac{u_{i+4}-4u_{i+3}+6u_{i+2}-4u_{i+1}+u_i}{\Delta X^4}$	Forward diff.	$O(\Delta X)$
	$\dfrac{-2u_{i+5}+11u_{i+4}-24u_{i+3}+26u_{i+2}-14u_{i+1}+3u_i}{\Delta X^4}$	"	$O(\Delta X^2)$
	$\dfrac{u_i-4u_{i-1}+6u_{i-2}-4u_{i-3}+u_{i-4}}{\Delta X^4}$	Backward diff.	$O(\Delta X)$
	$\dfrac{3u_i-14u_{i-1}+26u_{i-2}-24u_{i-3}+11u_{i-4}-2u_{i-5}}{\Delta X^4}$	"	$O(\Delta X^2)$
	$\dfrac{u_{i+2}-4u_{i+1}+6u_i-4u_{i-1}+u_{i-2}}{\Delta X^4}$	Centered diff.	$O(\Delta X^2)$

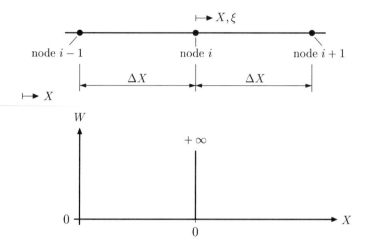

Fig. 1.3 Typical cell for the cell collocation method and appropriate weight function

$$N_{i-1} = \frac{1}{2}\xi(\xi - 1) \ , \ N_i = (1 - \xi)(1 + \xi) \ , \ N_{i+1} = \frac{1}{2}\xi(1 + \xi). \qquad (1.16)$$

The second-order derivative of the approximate function as given in Eq. (1.15) can be written under consideration of $\frac{\mathrm{d}\xi}{\mathrm{d}X} = \frac{1}{\Delta X}$ as:

$$\frac{\mathrm{d}^2 u}{\mathrm{d}X^2} = \frac{1}{\Delta X^2}\left(\frac{\mathrm{d}^2 N_{i-1}}{\mathrm{d}\xi^2}u_{i-1} + \frac{\mathrm{d}^2 N_i}{\mathrm{d}\xi^2}u_i + \frac{\mathrm{d}^2 N_{i+1}}{\mathrm{d}\xi^2}u_{i+1}\right)$$

$$= \frac{1}{\Delta X^2}(1u_{i-1} - 2u_i + 1u_{i+1}) \ . \qquad (1.17)$$

Thus, the residual

$$r = \frac{1}{\Delta X^2}(1u_{i-1} - 2u_i + 1u_{i+1}) \neq 0 \qquad (1.18)$$

can be used to formulate the collocation statement at node i as:

$$\int_{i-1}^{i+1} rW\mathrm{d}X = \int_{i-1}^{i+1} \frac{1}{\Delta X^2}(1u_{i-1} - 2u_i + 1u_{i+1})\,\delta(X - X_i) \overset{!}{=} 0 \ . \qquad (1.19)$$

Under consideration of the properties of the Dirac delta function [8], i.e.,

Fig. 1.4 Interpolation
functions for cell collocation
to derive a finite difference
scheme, adapted from [3]

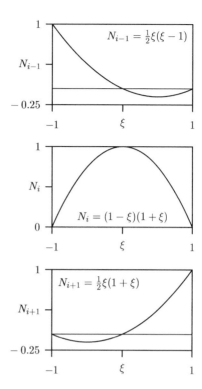

$$\int_{-\infty}^{\infty} \delta(X - X_k)\, dX = \int_{X_k-\varepsilon}^{X_k+\varepsilon} \delta(X - X_k)\, dX = 1\,, \tag{1.20}$$

$$\int_{-\infty}^{\infty} f(X)\delta(X - X_k)\, dX = \int_{X_k-\varepsilon}^{X_k+\varepsilon} f(X)\delta(X - X_k)\, dX = f(X_k)\,, \tag{1.21}$$

the last equation gives the statement

$$\frac{1}{\Delta X^2}\left(1u_{i-1} - 2u_i + 1u_{i+1}\right) = 0\,, \tag{1.22}$$

which is equivalent to the centered difference scheme as given in Table 1.1.

Instead of using the weighted residual method in the form of cell collocation, the method of collocation by subregions can be alternatively used to derive the finite difference method, cf. [3]. For this approach, the weight function is chosen as a step-type function as shown in Fig. 1.5.

Let us consider again a second-order derivative $\frac{d^2 u}{dX^2}$ for which the weighted residual statement can be written as:

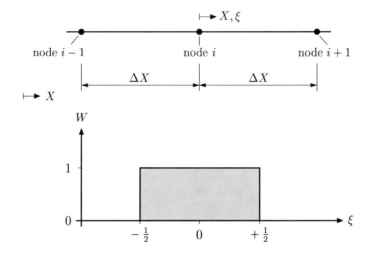

Fig. 1.5 Typical subregion for the collocation by subregion method and appropriate weight function, adapted from [3]

$$\int\limits_{-\Delta X/2}^{\Delta X/2} \frac{d^2 u}{dx^2} W \, dX = 0. \tag{1.23}$$

Integrating by parts gives the following expression

$$\int\limits_{-\Delta X/2}^{\Delta X/2} \frac{d^2 u}{dX^2} W \, dX = \left[\frac{du}{dX} W \right]_{-\Delta X/2}^{\Delta X/2} - \int\limits_{-\Delta X/2}^{\Delta X/2} \frac{du}{dX} \frac{dW}{dX} \, dX = 0, \tag{1.24}$$

or rearranged as:

$$\int\limits_{-\Delta X/2}^{\Delta X/2} \frac{du}{dX} \frac{dW}{dX} \, dX = \left[\frac{du}{dX} W \right]_{-\Delta X/2}^{\Delta X/2}. \tag{1.25}$$

It holds for the step function that $W = 1$ and $\frac{dW}{dX} = 0$ and the last equation can be simplified to:

$$0 = \frac{du}{dX}\bigg|_{+\Delta X/2} - \frac{du}{dX}\bigg|_{-\Delta X/2}. \tag{1.26}$$

Replacing the differentials 'd' by incremental differences, Eq. (1.26) can be written as

$$0 = \frac{u_{i+1} - u_i}{\Delta X} - \frac{u_i - u_{i-1}}{\Delta X} \, , \tag{1.27}$$

which corresponds to the centered difference scheme as given in Table 1.1.

1.1 Supplementary Problems

1.1 Forward difference approximation of the first order derivative

Derive the finite difference approximation for the forward difference scheme of the first-order derivative where the truncation error is of order ΔX^2. The final result is given in Table 1.1.

1.2 Centered difference approximation of the third order derivative

Derive the finite difference approximation for the centered difference scheme of the third-order derivative where the truncation error is of order ΔX^2 based on the expressions for the first- and second-order derivatives. These expressions and the final result are given in Table 1.1.

References

1. de Allen DN, G (1955) Relaxation methods. McGraw-Hill, New York
2. Bathe K-J (1996) Finite element procedures. Prentice-Hall, Upper Saddle River
3. Brebbia CA, Telles JFC, Wrobel LC (1984) Boundary element techniques: theory and applications. Springer, Berlin
4. Collatz L (1966) The numerical treatment of differential equations. Springer, Berlin
5. Forsythe GE, Wasow WR (1960) Finite-difference methods for partial differential equations. Wiley, New York
6. Lau PCM, Brebbia CA (1978) The cell collocation method in continuum mechanics. Int J Mech Sci 20:83–95
7. Mitchell AR, Griffiths DF (1980) The finite difference method in partial differential equations. Wiley, New York
8. Öchsner A (2014) Elasto-plasticity of frame structure elements: modeling and simulation of rods and beams. Springer, Berlin
9. Öchsner A, Merkel M (2018) One-dimensional finite elements: an introduction to the FE method. Springer, Cham
10. Öchsner A (2020) Computational statics and dynamics: an introduction based on the finite element method. Springer, Singapore
11. Southwell RV (1946) Relaxation methods in theoretical physics. Clarendon Press, Oxford

Chapter 2
Investigation of Rods in the Elastic Range

2.1 The Basics of a Rod

A rod is defined as a prismatic body whose axial dimension is much larger than its transverse dimensions [1, 3–5, 7, 9, 10]. This structural member is only loaded in the direction of the main body axes (X), see Fig. 2.1. As a result of this loading, the deformation occurs only along its main axis.

Derivations are restricted many times to the following simplifications:

- only applying to straight rods,
- displacements are (infinitesimally) small,
- strains are (infinitesimally) small, and
- the material is linear-elastic, i.e., constant Young's modulus E.

The three basic equations of continuum mechanics, i.e. the kinematics relationship, the constitutive law and the equilibrium equation, as well as their combination to the describing partial differential equation (PDE) are summarized in Table 2.1.

Alternative formulations of the partial differential equation for a rod are collected in Table 2.2. It should be noted here that some of the different cases given in Table 2.2 can be combined. The last case in Table 2.2 refers to the case of elastic embedding of a rod where the embedding modulus k has the unit of force per unit area.

Under the assumption of constant material ($E = $ const.) and geometric ($A = $ const.) properties, the first differential equation in Table 2.1 can be easily integrated twice for constant distributed load ($p_X = p_0 = $ const.) to obtain the general solution of the problem [6]:

$$u_X(X) = \frac{1}{EA}\left(-\frac{1}{2}p_0 X^2 + c_1 X + c_2\right), \tag{2.1}$$

where the two constants of integration c_i ($i = 1, 2$) must be determined based on the boundary conditions (see Table 2.3). The following equation for the internal normal force N_X was obtained based on one-time integration of the PDE and might be useful to determine some of the constants of integration:

© The Author(s), under exclusive license to Springer Nature Switzerland AG 2021
A. Öchsner, *Structural Mechanics with a Pen*,
https://doi.org/10.1007/978-3-030-65892-2_2

Fig. 2.1 Schematic representation of a continuum rod

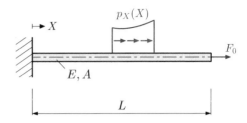

Table 2.1 Different formulations of the basic equations for a rod (X-axis along the principal rod axis), with $\mathcal{L}_1(\dots) = \frac{d(\dots)}{dX}$

Specific formulation	General formulation [1]
Kinematics	
$\varepsilon_X(X) = \frac{du_X(X)}{dX}$	$\varepsilon_X(X) = \mathcal{L}_1(u_X(X))$
Constitution	
$\sigma_X(X) = E\varepsilon_X(X)$	$\sigma_X(X) = C\varepsilon_X(X)$
Equilibrium	
$\frac{d\sigma_X(X)}{dX} + \frac{p_X(X)}{A} = 0$ or $\frac{dN_X(X)}{dX} + p_X(X) = 0$	$\mathcal{L}_1^{\mathrm{T}}(\sigma_X(X)) + b = 0$
PDE	
$\frac{d}{dX}\left(E(X)A(X)\frac{du_X}{dX}\right) + p_X(X) = 0$	$\mathcal{L}_1^{\mathrm{T}}(EA\mathcal{L}_1(u_X(X))) + p_X = 0$
$E(X)A(X)\frac{du_X}{dX} - N_X(X) = 0$	$EA\mathcal{L}_1(u_X(X)) - N_X = 0$

Table 2.2 Different formulations of the partial differential equation for a rod (X-axis: right facing)

Configuration	Partial differential equation
EA	$EA\frac{d^2u_X}{dX^2} = 0$
$E(X)A(X)$	$\frac{d}{dX}\left(E(X)A(X)\frac{du_X}{dX}\right) = 0$
$p_X(X)$	$EA\frac{d^2u_X}{dX^2} = -p_X(X)$
$k(X)$	$EA\frac{d^2u_X}{dX^2} = k(X)u_X$

Table 2.3 Different boundary conditions and corresponding reactions for a continuum rod (deformation occurs along the X-axis)

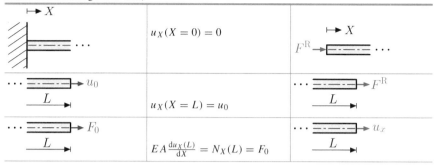

Fig. 2.2 Internal reactions for a continuum rod

$$N_{X(X)} = EA \frac{\mathrm{d}u_X(X)}{\mathrm{d}X} = -p_0 X + c_1 . \qquad (2.2)$$

The internal reactions in a rod become visible if one cuts—at an arbitrary location X—the member in two parts. As a result, two opposite oriented normal forces N_X can be indicated. Summing up the internal reactions from both parts must result in zero. Their positive direction is connected with the direction of the outward surface normal vector and the orientation of the positive X-axis, see Fig. 2.2.

Once the internal normal force N_X is known, the normal stress σ_X can be calculated:

$$\sigma_X(X) = \frac{N_X(X)}{A} . \qquad (2.3)$$

Application of Hooke's law (see Table 2.1) allows us to calculate the normal strain ε_X. Typical distributions of stress and strain in a rod element are shown in Fig. 2.3. It can be seen that both distributions are constant over the cross section.

2.2 Constant Material and Geometry Parameters

Let us consider in the following a loaded rod of length L as shown in Fig. 2.4. The tensile stiffness EA, i.e. the material and geometric parameters, are considered to be constant within this chapter. The left-hand end is fixed and the right-hand side is either loaded by a prescribed displacement u or a single force F. Along the length of the rod is a distributed load $p_X(X)$ acting.

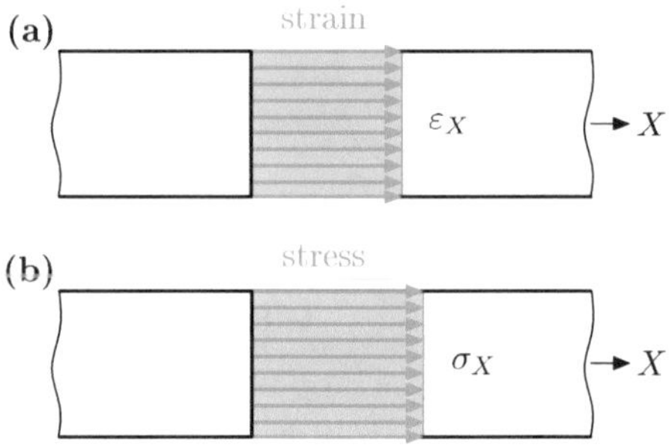

Fig. 2.3 Axially loaded rod:
a strain and **b** stress
distribution

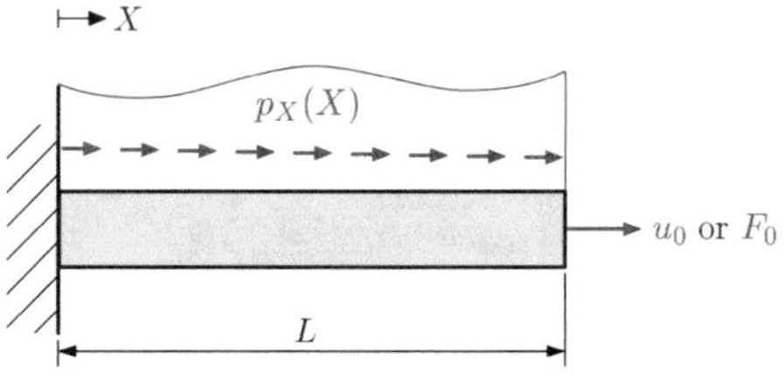

Fig. 2.4 Rod loaded with a
distributed load $p_X(X)$ and a
displacement boundary
condition u_0 or an end load
F_0

The problem is described according to Table 2.2 in the domain $X \in [0, L]$ by the
following partial differential equation

$$EA\frac{\mathrm{d}^2 u_X}{\mathrm{d}X^2} = -p_X(X) \tag{2.4}$$

and the boundary conditions

$$u_X(X = 0) = 0 \quad \text{and} \quad u_X(X = L) = u_0 \tag{2.5}$$

or

$$u_X(X = 0) = 0 \quad \text{and} \quad \left.\frac{\mathrm{d}u_X}{\mathrm{d}X}\right|_{X=L} = \frac{F_0}{EA}. \tag{2.6}$$

In a more formal way, Eq. (2.4) can be written as

$$EA\mathcal{L}\{u_X\} = -p_X(X), \tag{2.7}$$

where the differential operator is given by $\mathcal{L} = \frac{\mathrm{d}^2}{\mathrm{d}X^2}$. Taking from Table 1.1 the centered difference scheme of second-order accuracy, a finite difference approximation
of the partial differential equation (2.4) can be written for node i as[1]:

---

[1] To simplify the notation, the index X is omitted in the following derivations.

$$E A \left(\frac{u_{i+1} - 2u_i + u_{i-1}}{\Delta X^2} \right) = -p_i , \tag{2.8}$$

or as

$$\frac{E A}{\Delta X} (-u_{i+1} + 2u_i - u_{i-1}) = R_i , \tag{2.9}$$

where the equivalent nodal force R_i, resulting from a distributed load $p(X)$, is in general given for an *inner* node i as: $R_i = \int_{-\Delta x/2}^{\Delta x/2} p(\hat{x}) d\hat{x}$. In this integral, the local coordinate x has its origin at the location of node i. In the case of the boundary nodes, the equivalent nodal loads must be calculated as $R_1 = \int_0^{\Delta x/2} p(\hat{x}) d\hat{x}$ or $R_n = \int_{-\Delta x/2}^0 p(\hat{x}) d\hat{x}$ in order to completely distribute the entire load $p(X)$ to the nodes.

Let us assume for simplicity that the distributed load shown in Fig. 2.4 is zero ($p(X) = 0$) and introduce $n = 5$ nodes, i.e. two boundary nodes and three inner nodes. The evaluation of the finite difference approximation according to Eq. (2.9) at the inner nodes $i = 2, \ldots, 4$ gives:

$$\text{node 2:} \quad \frac{E A}{\Delta X} (-u_3 + 2u_2 - u_1) = 0 , \tag{2.10}$$

$$\text{node 3:} \quad \frac{E A}{\Delta X} (-u_4 + 2u_3 - u_2) = 0 , \tag{2.11}$$

$$\text{node 4:} \quad \frac{E A}{\Delta X} (-u_5 + 2u_4 - u_3) = 0 . \tag{2.12}$$

Considering at both ends displacement boundary conditions, i.e. Eq. (2.5), the system of equations (2.10) till (2.12) can be written with $u_1 = 0$ and $u_5 = u_0$ as

$$\text{node 2:} \quad \frac{E A}{\Delta X} (0 + 2u_2 - u_3) = 0 , \tag{2.13}$$

$$\text{node 3:} \quad \frac{E A}{\Delta X} (-u_2 + 2u_3 - u_4) = 0 , \tag{2.14}$$

$$\text{node 4:} \quad \frac{E A}{\Delta X} (-u_3 + 2u_4) = \frac{E A}{\Delta X} u_0 , \tag{2.15}$$

or in matrix notation as

$$\frac{E A}{\Delta X} \begin{bmatrix} 2 & -1 & 0 \\ -1 & 2 & -1 \\ 0 & -1 & 2 \end{bmatrix} \begin{bmatrix} u_2 \\ u_3 \\ u_4 \end{bmatrix} = \begin{bmatrix} 0 \\ 0 \\ \frac{E A}{\Delta X} u_0 \end{bmatrix} . \tag{2.16}$$

Fig. 2.5 Fictitious finite difference node $n + 1$ outside the rod, adapted from [2]

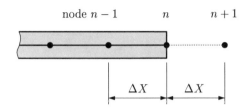

The last system of equations would be obtained the same way based on a finite element approximation with linear rod elements. Solution of the systems of equations given in Eq. (2.16) gives

$$
\begin{bmatrix} u_2 \\ u_3 \\ u_4 \end{bmatrix} = \begin{bmatrix} \frac{1}{4} u_0 \\ \frac{2}{4} u_0 \\ \frac{3}{4} u_0 \end{bmatrix} ,
\tag{2.17}
$$

which gives the same values as the analytical solution, i.e. linear increasing from 0 at $X = 0$ to u_0 at $X = L$.

The consideration of the force boundary condition at the right-hand end requires more efforts. Considering the equilibrium condition, i.e. $\frac{dN}{dX} = -p(X)$, together with the integrated form of the differential equation (see Table 2.1) gives the following expression for the internal normal force N as a function of the derivative of $u(X)$:

$$
EA \left(\frac{du(X)}{dX} \right)_i = N_i(X) .
\tag{2.18}
$$

To evaluate the gradient in Eq. (2.18), the different formulations given in Table 1.1 can be used. If the higher accurate centered difference scheme should be used, it is necessary to introduce a fictitious node outside the rod as shown in Fig. 2.5.

Then, the finite difference approximation for the boundary node 5 can be written as:

$$
\text{node 5:} \quad \frac{EA}{\Delta X} (-u_6 + 2u_5 - u_4) = 0 ,
\tag{2.19}
$$

and the gradient at the boundary node together with the force equilibrium, i.e. $N_5 = F_0$, reads as

$$
\left(\frac{du(X)}{dX} \right)_5 = \frac{u_6 - u_4}{2\Delta X} = \frac{F_0}{EA} ,
\tag{2.20}
$$

from which the expression for u_6 can be obtained as:

$$u_6 = u_4 + \frac{2F_0 \Delta X}{EA} \, . \tag{2.21}$$

This result can be introduced into Eq. (2.19) to receive the required expression to substitute u_5 into Eq. (2.12) by given values:

$$u_5 = u_4 + \frac{F_0 \Delta X}{EA} \, . \tag{2.22}$$

Thus, the final system of equations under consideration of the force boundary condition is given as:

$$\frac{EA}{\Delta X} \begin{bmatrix} 2 & -1 & 0 \\ -1 & 2 & -1 \\ 0 & -1 & 1 \end{bmatrix} \begin{bmatrix} u_2 \\ u_3 \\ u_4 \end{bmatrix} = \begin{bmatrix} 0 \\ 0 \\ F_0 \end{bmatrix} . \tag{2.23}$$

It should be mentioned here that the same result can be obtained in this case based on the finite element method if the equation for node 5 is eliminated. Solution of this system of equations gives

$$\begin{bmatrix} u_2 \\ u_3 \\ u_4 \end{bmatrix} = \begin{bmatrix} \frac{\Delta X F_0}{EA} \\ \frac{2\Delta X F_0}{EA} \\ \frac{3\Delta X F_0}{EA} \end{bmatrix} , \tag{2.24}$$

which gives the same result as the analytical solution under consideration of $\Delta X = \frac{L}{4}$, see [6]. Alternatively, we may write the system of equations under consideration of the displacement at node 5 as:

$$\frac{EA}{\Delta X} \begin{bmatrix} 2 & -1 & 0 & 0 \\ -1 & 2 & -1 & 0 \\ 0 & -1 & 2 & -1 \\ 0 & 0 & -1 & 1 \end{bmatrix} \begin{bmatrix} u_2 \\ u_3 \\ u_4 \\ u_5 \end{bmatrix} = \begin{bmatrix} 0 \\ 0 \\ 0 \\ F_0 \end{bmatrix} , \tag{2.25}$$

or solved for the nodal unknowns:

$$\begin{bmatrix} u_2 \\ u_3 \\ u_4 \\ u_5 \end{bmatrix} = \begin{bmatrix} \frac{\Delta X F_0}{EA} \\ \frac{2\Delta X F_0}{EA} \\ \frac{3\Delta X F_0}{EA} \\ \frac{4\Delta X F_0}{EA} \end{bmatrix} . \tag{2.26}$$

Substituting $\Delta X = \frac{L}{4}$ gives the exact analytical solution as presented in [6].

An alternative approach can be based on the first-order differential equation in the formulation of the internal normal force (see the second formulation of the PDE in Table 2.1), i.e. for constant material (E) and geometric (A) properties:

$$EA\frac{du_X}{dX} = N_X(X). \tag{2.27}$$

This means that the function of the internal normal force distribution ($N_X(X)$) must be determined. Taking from Table 1.1 the centered difference scheme of second-order accuracy, a finite difference approximation of the partial differential equation (2.27) can be written for node i as follows:

$$\frac{EA}{2\Delta X}(u_{i+1} - u_{i-1}) = N_X(X_i). \tag{2.28}$$

2.3 Varying Material and Geometry Parameters

Let us consider in the following the case that the tensile stiffness is a function of the Cartesian coordinate X. Thus, the generalized problem shown in Fig. 2.4 can be described in the domain $X \in [0, L]$ by the following partial differential equation

$$\frac{d}{dX}\left(E(X)A(X)\frac{du}{dX}\right) = -p(X) \tag{2.29}$$

and the boundary conditions

$$u(X = 0) = 0 \text{ and } u(X = L) = u_0 \tag{2.30}$$

or

$$u(X = 0) = 0 \text{ and } \left.\frac{du}{dX}\right|_{X=L} = \frac{F}{E(L)A(L)}. \tag{2.31}$$

The product of the varying modulus and cross section can be combined in an auxiliary function

$$k(X) = E(X)A(X), \tag{2.32}$$

and the differential equation (2.33) reads in a more general notation as:

$$\frac{d}{dX}\left(k(X)\frac{du}{dX}\right) = -p(X). \tag{2.33}$$

A finite difference approximation of this equation can be introduced in different ways. Let us first neglect for simplicity the distributed load ($p(X) = 0$) since the treatment of a distributed load is discussed in Sect. 2.2.

The first approach is based on the introduction of an auxiliary function $v(x)$ of the form:

$$v(X) = k(X)\frac{du}{dX} \, . \tag{2.34}$$

Thus, the differential equation (2.33) can be expressed in a much more simpler way under the simplification that the distributed load is neglected as:

$$\frac{dv(X)}{dX} = 0 \, . \tag{2.35}$$

The first task is now to approximate the first-order derivative $\frac{dv}{dX}$. To this end, let us consider the centered difference expression from Table 1.1 and replace ΔX by $\frac{\Delta X}{2}$. Thus, the centered difference approximation is given by

$$\left(\frac{dv(X)}{dX}\right)_i = \frac{v_{i+\frac{1}{2}} - v_{i-\frac{1}{2}}}{\Delta X} \, , \tag{2.36}$$

where, for example, the notation '$i + \frac{1}{2}$' refers to the value of the function v in the middle of node i and $i + 1$, cf. Fig. 2.6.

Based on the definition of the auxiliary function v in Eq. (2.34), the functional values of v in Eq. (2.36) at the intermediate locations can be stated based on centered difference approximations of second-order accuracy as ($\Delta X \rightarrow \Delta X/2$):

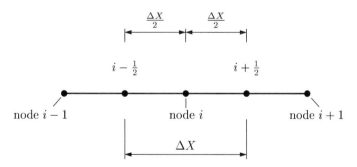

Fig. 2.6 Definition of the intermediate position $i - \frac{1}{2}$ and $i + \frac{1}{2}$

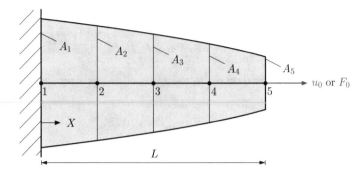

Fig. 2.7 Rod with linear varying cross-section area discretized by five nodes in a finite difference approach

$$v_{i+\frac{1}{2}} = k_{i+\frac{1}{2}} \left(\frac{du}{dX} \right)_{i+\frac{1}{2}} = k_{i+\frac{1}{2}} \frac{u_{i+1} - u_i}{\Delta X}, \tag{2.37}$$

$$v_{i-\frac{1}{2}} = k_{i-\frac{1}{2}} \left(\frac{du}{dX} \right)_{i-\frac{1}{2}} = k_{i-\frac{1}{2}} \frac{u_i - u_{i-1}}{\Delta X}. \tag{2.38}$$

Thus, the finite difference approximation of the problem given in Eq. (2.35) can be written for node i as:

$$\frac{1}{\Delta X} \left(-k_{i-\frac{1}{2}} u_{i-1} + (k_{i-\frac{1}{2}} + k_{i+\frac{1}{2}}) u_i - k_{i+\frac{1}{2}} u_{i+1} \right) = 0. \tag{2.39}$$

It should be noted here that the last equation was multiplied by ΔX. This is in general required to obtain from distributed loads the equivalent nodal force R_i, cf. Sect. 2.2. In addition, the last equation was multiplied by '-1' to make the comparison with the finite element approach easier.

In order to simplify the explanations of the different approaches and possibilities, let us consider in the following a rod with linear varying cross-sectional area as shown in Fig. 2.7. This structure should be discretized by five nodes and either a displacement u_0 or a force F_0 should be prescribed at the right-hand end.

Evaluation of the finite difference scheme given in Eq. (2.39) for the inner nodes shown in Fig. 2.7 gives:

$$\text{node 2:} \frac{E}{\Delta X} \left(-\frac{A_1 + A_2}{2} u_1 + \left(\frac{A_1 + A_2}{2} + \frac{A_2 + A_3}{2} \right) u_2 - \frac{A_2 + A_3}{2} u_3 \right) = 0, \tag{2.40}$$

$$\text{node 3:} \frac{E}{\Delta X} \left(-\frac{A_2 + A_3}{2} u_2 + \left(\frac{A_2 + A_3}{2} + \frac{A_3 + A_4}{2} \right) u_3 - \frac{A_3 + A_4}{2} u_4 \right) = 0, \tag{2.41}$$

$$\text{node 4:} \frac{E}{\Delta X} \left(-\frac{A_3 + A_4}{2} u_3 + \left(\frac{A_3 + A_4}{2} + \frac{A_4 + A_5}{2} \right) u_4 - \frac{A_4 + A_5}{2} u_5 \right) = 0.$$

$$(2.42)$$

The consideration of the displacement boundary conditions according to Eq. (2.30), i.e. $u_1 = 0$ and $u_5 = u_0$, in the three last equations and rearranging the set of equations in matrix form results in:

$$\frac{E}{2\Delta X} \begin{bmatrix} A_1 + 2A_2 + A_3 & -(A_2 + A_3) & 0 \\ -(A_2 + A_3) & A_2 + 2A_3 + A_4 & -(A_3 + A_4) \\ 0 & -(A_3 + A_4) & A_3 + 2A_4 + A_5 \end{bmatrix} \begin{bmatrix} u_2 \\ u_3 \\ u_4 \end{bmatrix} =$$

$$\begin{bmatrix} 0 \\ 0 \\ \frac{E}{2\Delta X}(A_4 + A_5)u_0 \end{bmatrix}. \qquad (2.43)$$

The same system of equations would be obtained based on a finite element approach with linear elements, see [8].

If the boundary condition at the right-hand end is given as a single force F_0, one may follow first of all the idea based on a fictitious node $n + 1$ outside the structure as shown in the previous section in Fig. 2.5. The evaluation of node 5 based on the general scheme (2.39) gives

$$\text{node 5:} \frac{E}{\Delta X} \left(-\frac{A_4 + A_5}{2} u_4 + \left(\frac{A_4 + A_5}{2} + \frac{A_5 + A_6}{2} \right) u_5 - \frac{A_5 + A_6}{2} u_6 \right) = 0.$$

$$(2.44)$$

If the centered difference scheme from Table 1.1 is used, the gradient at the boundary node together with the force equilibrium, i.e. $N_5 = F_5$, reads as

$$\left(\frac{du}{dX} \right)_5 = \frac{u_6 - u_4}{2\Delta X} = \frac{F_0}{EA_5}. \qquad (2.45)$$

The last equation can be rearranged for u_6 and this relationship can be introduced into the evaluation of node 5 according to Eq. (2.44). This gives finally an expression for the displacement u_5 in the form:

$$u_5 = u_4 + \frac{2F_0 \Delta X}{E} \frac{A_5 + A_6}{A_5(A_4 + 2A_5 + A_6)}. \qquad (2.46)$$

The last equation still contains the fictitious area A_6 and might be not the best approach to solve the problem.

An alternative approach can be based on the backward difference scheme as given in Table 1.1. Then, the gradient at the boundary node together with the force equilibrium reads as:

$$\left(\frac{du}{dX}\right)_5 = \frac{u_5 - u_4}{\Delta X} = \frac{F_0}{E A_5},\tag{2.47}$$

or rearranged for u_5 as:

$$u_5 = u_4 + \frac{F_0 \Delta X}{E A_5}.\tag{2.48}$$

The last equation can be introduced into the evaluation of node 4 according to Eq. (2.42) to finally result in:

$$\frac{E}{2\Delta X}(-(A_3 + A_4)u_3 + (A_3 + A_4)u_4) = \frac{A_4 + A_5}{2A_5} F_0.\tag{2.49}$$

Thus, the final system of equations is given by Eqs. (2.40) and (2.41) under consideration of the boundary condition at node 1 ($u_1 = 0$) and Eq. (2.49) which considers the force boundary condition at node 5. This systems of equations reads in matrix from as:

$$\frac{E}{2\Delta X}\begin{bmatrix} A_1 + 2A_2 + A_3 & -(A_2 + A_3) & 0 \\ -(A_2 + A_3) & A_2 + 2A_3 + A_4 & -(A_3 + A_4) \\ 0 & -(A_3 + A_4) & A_3 + A_4 \end{bmatrix}\begin{bmatrix} u_2 \\ u_3 \\ u_4 \end{bmatrix} = \begin{bmatrix} 0 \\ 0 \\ \frac{A_4 + A_5}{2A_5} F_0 \end{bmatrix}.\tag{2.50}$$

The left-hand side of the last equation is identical to the finite element approach but the load vector differs from the finite element solution where only F would appear.

An alternative approach can be based on the idea that the first order derivative is evaluated in the middle of node $i - 1$ and i due to a centered difference scheme, cf. Fig. 2.8a.

The corresponding centered difference scheme is given by

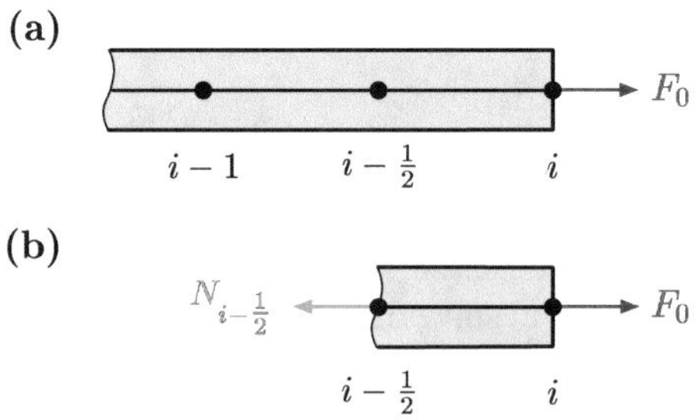

Fig. 2.8 Definition of a centered difference scheme at the position $i - \frac{1}{2}$. Note that the distance between node $i - 1$ and i is equal to ΔX

(a)

$i - 1$ $\quad i - \frac{1}{2}$ $\quad i$ $\qquad F_0$

(b)

$N_{i-\frac{1}{2}}$ $\qquad F_0$

$i - \frac{1}{2}$ $\quad i$

$$\left(\frac{du}{dX}\right)_{i-\frac{1}{2}} = \frac{u_i - u_{i-1}}{\Delta X} = \frac{N_{i-\frac{1}{2}}}{k_{i-\frac{1}{2}}}, \tag{2.51}$$

whereas it follows from the force equilibrium (cf. Fig. 2.8b) that $N_{i-\frac{1}{2}} = F_0$ as along as there is no distributed load involved in the considered section of the rod. Applying the node numbering given in Fig. 2.7, Eq. (2.51) can be rearranged to obtain the following expression:

$$u_5 = u_4 + \frac{2F_0\Delta X}{E(A_4 + A_5)}, \tag{2.52}$$

which can be introduced into Eq. (2.42) to finally obtain:

$$\frac{E}{2\Delta X}(-(A_3 + A_4)u_3 + (A_3 + A_4)u_4) = F_0. \tag{2.53}$$

The last equation can be combined with Eqs. (2.40) and (2.41) to obtain the systems of equations as:

$$\frac{E}{2\Delta X}\begin{bmatrix} A_1 + 2A_2 + A_3 & -(A_2 + A_3) & 0 \\ -(A_2 + A_3) & A_2 + 2A_3 + A_4 & -(A_3 + A_4) \\ 0 & -(A_3 + A_4) & A_3 + A_4 \end{bmatrix}\begin{bmatrix} u_2 \\ u_3 \\ u_4 \end{bmatrix} = \begin{bmatrix} 0 \\ 0 \\ F_0 \end{bmatrix}. \tag{2.54}$$

This formulation is identical to the finite element approach.

To investigate the difference between the system of equations given in Eqs. (2.50) and (2.54), let us consider in the following different ratios a between the area A_5 and A_1, cf. Fig. 2.7. The comparison with the analytical solution, see Ref. [6], will provide some understanding which is the better approach.

From Table 2.4 where the numerical errors between the different implementations of the force boundary condition and the exact solution are summarized, it can be concluded that the approach based on Eq. (2.54) gives a much more accurate approximation of the problem under consideration. This is a direct result of that fact that Eq. (2.54) is derived under consideration of a centered difference scheme (error $\sim O(\Delta X^2)$), whereas Eq. (2.50) is based on a backward difference scheme where the error is of order $O(\Delta X)$. Table 2.4 contains no numerical errors for the coordinate $X = 0$ since the result is exact (boundary condition) and the fraction would give $\frac{0}{0}$. Furthermore, it can be seen that the error is decreasing for increasing ratio A_5/A_1. For the limiting case $A_5 = A_1$, i.e. constant cross-sectional area, the solution would be exact.

A different approach than the presented strategies can be based on the product rule of differential calculus. Let us assume for simplicity that the Young's modulus in the differential equation according to Eq. (2.33) is constant and consider the product of the functions $A(X)$ and $\frac{du}{dX}$. Then, the differential of the product is given by

Table 2.4 Relative error, $\frac{u_{\text{approx}} - u_{\text{exact}}}{u_{\text{exact}}} \times 100$, in percent for different implementations of a force boundary condition

Eq.	Node 1	Node 2	Node 3	Node 4	Node 5
	$X = 0$	$X = \frac{L}{4}$	$X = \frac{2L}{4}$	$X = \frac{4L}{4}$	$X = L$
$a = \frac{A_5}{A_1} = 0.1$					
(2.50)	–	111.3569	110.8318	109.3803	48.1861
(2.54)	–	−0.5379	−0.7850	−1.4681	−5.9208
$a = \frac{A_5}{A_1} = 0.2$					
(2.50)	–	49.3807	49.1516	48.6324	26.0422
(2.54)	–	−0.4129	−0.5656	−0.9117	−2.1644
$a = \frac{A_5}{A_1} = 0.3$					
(2.50)	–	28.7698	28.6528	28.4275	16.9195
(2.54)	–	−0.3073	−0.3979	−0.5723	−1.0116

$$E\left(\frac{dA(X)}{dX}\frac{du}{dX} + A(X)\frac{d^2u}{dX^2}\right) = 0. \qquad (2.55)$$

Introduction of centered difference schemes for the first- and second-order derivative according to Table 1.1 gives:

$$E\left(\left(\frac{dA}{dX}\right)_i \frac{u_{i+1} - u_{i-1}}{2\Delta X} + A_i \frac{u_{i+1} - 2u_i + u_{i-1}}{\Delta X^2}\right) = 0. \qquad (2.56)$$

Let us assume as in the previous derivations a linear varying cross-sectional area. Thus, the gradient of the cross-sectional area function can be introduced similar to the centered difference scheme[2] as

$$\frac{dA}{dX} = \frac{A_{i+1} - A_{i-1}}{2\Delta X}, \qquad (2.57)$$

which can be introduced into Eq. (2.56) to finally obtain the finite difference scheme:

$$\frac{E}{\Delta X}\left(-\left[\frac{1}{4}A_{i-1} + A_i - \frac{1}{4}A_{i+1}\right]u_{i-1} + 2A_iu_i \right.$$
$$\left. -\left[-\frac{1}{4}A_{i-1} + A_i + \frac{1}{4}A_{i+1}\right]u_{i+1}\right). \qquad (2.58)$$

---

[2] At this stage, many different ways are possible on how to introduce this gradient. A formulation similar to the centered difference scheme seems the natural choice and other approaches will be not discussed here.

It should be noted here that the last equation is multiplied by ΔX and (-1). Evaluation of this finite difference scheme for the inner nodes ($i = 2, 3, 4$) shown in Fig. 2.7 gives:

$$\frac{E}{\Delta X}\left(-\left[\frac{1}{4}A_1 + A_2 - \frac{1}{4}A_3\right]u_1 + 2A_2u_2 - \left[-\frac{1}{4}A_1 + A_2 + \frac{1}{4}A_3\right]u_3\right) = 0,$$
$$(2.59)$$

$$\frac{E}{\Delta X}\left(-\left[\frac{1}{4}A_2 + A_3 - \frac{1}{4}A_4\right]u_2 + 2A_3u_3 - \left[-\frac{1}{4}A_2 + A_3 + \frac{1}{4}A_4\right]u_4\right) = 0,$$
$$(2.60)$$

$$\frac{E}{\Delta X}\left(-\left[\frac{1}{4}A_3 + A_4 - \frac{1}{4}A_5\right]u_3 + 2A_4u_4 - \left[-\frac{1}{4}A_3 + A_4 + \frac{1}{4}A_5\right]u_4\right) = 0.$$
$$(2.61)$$

The force boundary condition can be introduced based on the idea of the centered difference scheme as given in Eq. (2.52). Based on this expression, Eq. (2.61) can be rewritten to obtain the following expression:

$$\frac{E}{\Delta X}\left(-\left[\frac{1}{4}A_3 + A_4 - \frac{1}{4}A_5\right]u_3 - \left[-\frac{1}{4}A_3 - A_4 + \frac{1}{4}A_5\right]u_4\right) =$$
$$F_0\frac{-\frac{1}{2}A_3 + 2A_4 + \frac{1}{2}A_5}{A_4 + A_5}. \quad (2.62)$$

Thus, the final system of equations for this approach, which is based on the product rule of differential calculus, is given by the following matrix scheme:

$$\frac{E}{2\Delta X}\begin{bmatrix} 4A_2 & \frac{A_1}{2} - 2A_2 - \frac{A_3}{2} & 0 \\ -\frac{A_2}{2} - 2A_3 - \frac{A_4}{2} & 4A_3 & \frac{A_2}{2} - 2A_3 - \frac{A_4}{2} \\ 0 & -\frac{A_3}{2} - 2A_4 + \frac{A_5}{2} & \frac{A_3}{2} + 2A_4 - \frac{A_5}{2} \end{bmatrix}\begin{bmatrix} u_2 \\ u_3 \\ u_4 \end{bmatrix} =$$
$$\begin{bmatrix} 0 \\ 0 \\ F_0\frac{-\frac{1}{2}A_3 + 2A_4 + \frac{1}{2}A_5}{A_4 + A_5} \end{bmatrix}. \quad (2.63)$$

The evaluation of the cases presented in Table 2.4 reveals that this derivation gives identical results as Eq. (2.54), i.e. an identical solution as a finite element approach.

2.4 Solved Problems

2.1 Finite difference approximation of a cantilevered rod with a single force based on five domain nodes

Given is a cantilevered rod of length L with constant tensile stiffness EA as shown in Fig. 2.9. The rod is loaded by a single force F_0 at its right-hand boundary. Use five domain nodes of equidistant spacing, i.e. $\Delta X = \frac{L}{4}$, for the finite difference approximation. Use only finite difference approximations of second-order accuracy for the nodal evaluations and boundary conditions. Perform the evaluations (a) starting from the second-order partial differential equation and (b) as an alternative starting from the first-order partial differential equation. Determine

- the horizontal displacements at the nodes,
- the analytical solution at $X = L$ and
- calculate the relative error between the analytical and finite difference solution at $X = L$.
- Derive a general FD scheme for n nodes ($n > 5$).

2.1 Solution

The finite difference discretization of the simply cantilevered rod is shown in Fig. 2.10.

(a) Let us start with the second-order partial differential equation as given in Table 2.1. Application of a centered difference approximations of second-order accuracy (see Table 1.1), one can state the following FD scheme for node i:

$$\frac{EA}{\Delta X^2} (-u_{i-1} + 2u_i - u_{i+1}) = R_i \overset{p_{X=0}}{=} 0. \tag{2.64}$$

Since we have four unknown nodal displacements ($u_2, \ldots, u_5$), we must state four equations to solve the problem. Thus, we could write the FD approximation in Eq. (2.64) for the four nodes $i = 1, \ldots, 4$ or for the four nodes $i = 2, \ldots, 5$. Both approaches are possible. Let us write the FD scheme in the following for nodes

Fig. 2.9 Cantilevered rod loaded by a single force

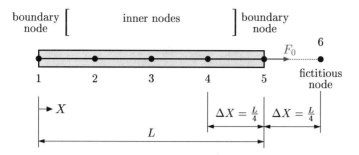

Fig. 2.10 Finite difference discretization of the cantilevered rod

$i = 2, \ldots, 5$ and introduce by doing so a fictitious node at the right-hand boundary (see Fig. 2.10):

$$\text{node 2: } \frac{EA}{\Delta X^2} (-u_1 + 2u_2 - u_3) = 0, \tag{2.65}$$

$$\text{node 3: } \frac{EA}{\Delta X^2} (-u_2 + 2u_3 - u_4) = 0, \tag{2.66}$$

$$\text{node 4: } \frac{EA}{\Delta X^2} (-u_3 + 2u_4 - u_5) = 0, \tag{2.67}$$

$$\text{node 5: } \frac{EA}{\Delta X^2} (-u_4 + 2u_5 - u_6) = 0. \tag{2.68}$$

It must be emphasized here that the right-hand side of Eq. (2.68) is zero and the external force F_0 should *not* be considered here. Furthermore, it should be noted that Eqs. (2.66)–(2.67), i.e., the equations with the gray background, are not affected by any fictitious nodes or nodes where displacements are imposed to the structure. These equations will help us later to construct a scheme for a larger number of nodes ($n > 5$). The horizontal displacement is zero at the left-hand end and it can be immediately concluded that $u_1 = 0$. Equation (2.68) contains still the displacement of the fictitious node 6 and a balance between the internal normal force (N_X) and the external load (F_0) at node 5 allows to derive a relationship to eliminate u_6 and to introduce the external load F_0:

$$EA \frac{du_X}{dX}\bigg|_5 = \frac{EA}{2\Delta X} (u_6 - u_4) = N_5 = F_0, \tag{2.69}$$

or rearranged for the displacement of the fictitious node:

$$u_6 = u_4 + \frac{2\Delta X F_0}{EA}.$$ (2.70)

Introducing the last relation in Eq. (2.68), the following matrix scheme can be stated:

$$\begin{bmatrix} 2 & -1 & 0 & 0 \\ -1 & 2 & -1 & 0 \\ 0 & -1 & 2 & -1 \\ 0 & 0 & -2 & 2 \end{bmatrix} \begin{bmatrix} u_2 \\ u_3 \\ u_4 \\ u_5 \end{bmatrix} = \frac{\Delta X F_0}{EA} \begin{bmatrix} 0 \\ 0 \\ 0 \\ 2 \end{bmatrix}.$$ (2.71)

The solution of this linear system of equations gives the unknown nodal values as:

$$\begin{bmatrix} u_2 \\ u_3 \\ u_4 \\ u_5 \end{bmatrix} = \frac{F_0 L}{EA} \begin{bmatrix} \frac{1}{4} \\ \frac{2}{4} \\ \frac{3}{4} \\ \frac{4}{4} \end{bmatrix},$$ (2.72)

and the relative error at the right-hand end of the rod is obtained as:

$$\text{relative error} = \frac{1 - 1}{1} \times 100 = 0.0\%,$$ (2.73)

this means that the numerical FD solution is identical with the analytical solution in this case, see [6]. From the above calculations, it is easy to derive a general scheme for n nodes ($n > 5$). In generalization of Eq. (2.71), the following scheme can be proposed:

$$\begin{bmatrix} 2 & -1 & 0 & 0 & \cdots & 0 \\ -1 & 2 & -1 & 0 & \cdots & 0 \\ 0 & -1 & 2 & -1 & \cdots & 0 \\ \vdots & \cdots & & & \cdots & \vdots \\ 0 & \cdots & 0 & -1 & 2 & -1 \\ 0 & \cdots & 0 & 0 & -2 & 2 \end{bmatrix} \begin{bmatrix} u_2 \\ u_3 \\ u_4 \\ \vdots \\ u_{n-1} \\ u_n \end{bmatrix} = -\frac{\Delta X F_0}{EA} \begin{bmatrix} 0 \\ 0 \\ 0 \\ \vdots \\ 0 \\ 2 \end{bmatrix},$$ (2.74)

where $\Delta X = \frac{L}{n-1}$ for equidistant spacing.

(b) Let us now focus on the first-order partial differential equation as given in Table 2.1. Application of a centered difference approximations of second-order accuracy (see Table 1.1), one can state the following FD scheme for node i:

$$\frac{EA}{2\Delta X} (u_{i+1} - u_{i-1}) = N_X = F_0.$$ (2.75)

Let us write the FD scheme in the following for nodes $i = 2, \ldots, 5$ and introduce by doing so a fictitious node at the right-hand boundary (see Fig. 2.10):

$$\text{node 2:} \quad \frac{EA}{2\Delta X}(u_3 - u_1) = F_0, \tag{2.76}$$

$$\text{node 3:} \quad \frac{EA}{2\Delta X}(u_4 - u_2) = F_0, \tag{2.77}$$

$$\text{node 4:} \quad \frac{EA}{2\Delta X}(u_5 - u_3) = F_0, \tag{2.78}$$

$$\text{node 5:} \quad \frac{EA}{2\Delta X}(u_6 - u_4) = F_0. \tag{2.79}$$

However, it turns out that the displacement of the fictitious node (u_6) cannot be replaced in Eq. (2.79) since we stated already the moment equation. Thus, we apply a backward difference scheme (see Table 1.1) to obtain a new fifth equation:

$$\text{node 5:} \quad \frac{EA}{2\Delta X}(3u_5 - 4u_4 + u_3) = F_0. \tag{2.80}$$

From the above derivations, the following matrix scheme can be stated:

$$\begin{bmatrix} 0 & 1 & 0 & 0 \\ -1 & 0 & 1 & 0 \\ 0 & -1 & 0 & 1 \\ 0 & 1 & -4 & 3 \end{bmatrix} \begin{bmatrix} u_2 \\ u_3 \\ u_4 \\ u_5 \end{bmatrix} = \frac{2\Delta X F_0}{EA} \begin{bmatrix} 1 \\ 1 \\ 1 \\ 1 \end{bmatrix}. \tag{2.81}$$

The solution of this linear system of equations gives the unknown nodal values as:

$$\begin{bmatrix} u_2 \\ u_3 \\ u_4 \\ u_5 \end{bmatrix} = \frac{F_0 L}{EA} \begin{bmatrix} \frac{1}{4} \\ \frac{2}{4} \\ \frac{3}{4} \\ \frac{4}{4} \end{bmatrix}, \tag{2.82}$$

and the relative error at the right-hand end of the rod is obtained as:

$$\text{relative error} = \frac{1 - 1}{1} \times 100 = 0.0\%, \tag{2.83}$$

Fig. 2.11 Stepped
cantilevered rod loaded by a
single force

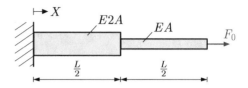

this means that the numerical FD solution is identical with the analytical solution in
this case, see [6]. From the above calculations, it is easy to derive a general scheme
for n nodes $(n > 5)$. In generalization of Eq. (2.81), the following scheme can be
proposed:

$$
\begin{bmatrix}
0 & 1 & 0 & 0 & \cdots & 0 \\
-1 & 0 & 1 & 0 & \cdots & 0 \\
0 & -1 & 0 & 1 & \cdots & 0 \\
\vdots & & \cdots & & \cdots & \vdots \\
0 & \cdots & 0 & -1 & 0 & 1 \\
0 & \cdots & 0 & 1 & -4 & 3
\end{bmatrix}
\begin{bmatrix}
u_2 \\
u_3 \\
u_4 \\
\vdots \\
u_{n-1} \\
u_n
\end{bmatrix}
= \frac{2\Delta X F_0}{EA}
\begin{bmatrix}
1 \\
1 \\
1 \\
\vdots \\
1 \\
1
\end{bmatrix},
\qquad (2.84)
$$

where $\Delta X = \frac{L}{n-1}$ for equidistant spacing.

2.2 Finite difference approximation of a stepped cantilevered rod with a single force based on five domain nodes

Given is a stepped rod of length L with a tensile stiffness of $E(2A)$ in the range $0 \le X \le L/2$ and a value of EA in the range $L/2 \le X \le L$ as shown in Fig. 2.11. The rod is loaded by a single force F_0 at its right-hand boundary. Use five domain nodes of equidistant spacing, i.e. $\Delta X = \frac{L}{4}$, for the finite difference approximation. Use only finite difference approximations of second-order accuracy for the nodal evaluations and boundary conditions. Perform the evaluations (a) starting from Eq. (2.39) and as an alternative considering (b) Eq. (2.51) for node 5, and (c) starting from Eq. (2.56). Determine the horizontal displacements at the nodes.

2.2 Solution

The finite difference discretization of the simply cantilevered rod is shown in Fig. 2.12.

(a) Starting from Eq. (2.39), the following FD scheme can be indicated for node i:

$$
\frac{E}{\Delta X}\left(-A_{i-\frac{1}{2}}u_{i-1} + (A_{i-\frac{1}{2}} + A_{i+\frac{1}{2}})u_i - A_{i+\frac{1}{2}}u_{i+1}\right) = 0. \qquad (2.85)
$$

Let us write the FD scheme in the following for nodes $i = 2, \ldots, 5$ and introduce by doing so a fictitious node at the right-hand boundary (see Fig. 2.12):

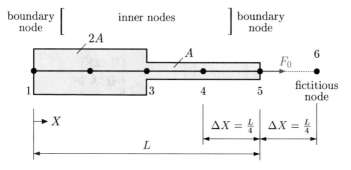

Fig. 2.12 Finite difference discretization of the stepped rod

$$\text{node 2: } \frac{E}{\Delta X^2}(-2Au_1 + (2A + 2A)u_2 - 2Au_3) = 0\,, \tag{2.86}$$

$$\text{node 3: } \frac{E}{\Delta X^2}(-2Au_2 + (2A + A)u_3 - Au_4) = 0\,, \tag{2.87}$$

$$\text{node 4: } \frac{E}{\Delta X^2}(-Au_3 + (A + A)u_4 - Au_5) = 0\,, \tag{2.88}$$

$$\text{node 5: } \frac{E}{\Delta X^2}(-Au_4 + (A + A)u_5 - Au_6) = 0\,. \tag{2.89}$$

The horizontal displacement is zero at the left-hand end and it can be immediately concluded that $u_1 = 0$. Equation (2.89) contains still the displacement of the fictitious node 6 and a balance between the internal normal force (N_X) and the external load (F_0) at node 5 allows to derive a relationship to eliminate u_6 (see Example 2.1 for details):

$$u_6 = u_4 + \frac{2\Delta X F_0}{E_5 A_5}\,. \tag{2.90}$$

Introducing the last relation in Eq. (2.89), the following matrix scheme can be stated:

$$\begin{bmatrix} 4A & -2A & 0 & 0 \\ -2A & 3A & -A & 0 \\ 0 & -A & 2A & -A \\ 0 & 0 & -2A & 2A \end{bmatrix} \begin{bmatrix} u_2 \\ u_3 \\ u_4 \\ u_5 \end{bmatrix} = \frac{\Delta X F_0}{E} \begin{bmatrix} 0 \\ 0 \\ 0 \\ 2 \end{bmatrix}\,, \tag{2.91}$$

or

$$\begin{bmatrix} 4 & -2 & 0 & 0 \\ -2 & 3 & -1 & 0 \\ 0 & -1 & 2 & -1 \\ 0 & 0 & -2 & 2 \end{bmatrix} \begin{bmatrix} u_2 \\ u_3 \\ u_4 \\ u_5 \end{bmatrix} = \frac{\Delta X F_0}{EA} \begin{bmatrix} 0 \\ 0 \\ 0 \\ 2 \end{bmatrix}\,. \tag{2.92}$$

The solution of this linear system of equations gives the unknown nodal values as:

$$\begin{bmatrix} u_2 \\ u_3 \\ u_4 \\ u_5 \end{bmatrix} = \frac{F_0 L}{E A} \begin{bmatrix} \frac{1}{8} \\ \frac{1}{4} \\ \frac{1}{2} \\ \frac{3}{4} \end{bmatrix} = \frac{F_0 L}{E A} \begin{bmatrix} 0.125 \\ 0.25 \\ 0.5 \\ 0.75 \end{bmatrix}.$$

(2.93)

This solution is identical to the analytical solution.

(b) Considering Eq. (2.51) for node 5, the following FD scheme can be indicated for this node:

$$k_{5-\frac{1}{2}} \frac{u_5 - u_4}{\Delta X} = N_{5-\frac{1}{2}} = F_0 ,$$

(2.94)

or:

$$\text{node 5: } u_5 - u_5 - \frac{\Delta X F_0}{E A} .$$

(2.95)

Thus, the following matrix scheme can be stated:

$$\begin{bmatrix} 4 & -2 & 0 & 0 \\ -2 & 3 & -1 & 0 \\ 0 & -1 & 2 & -1 \\ 0 & 0 & -1 & 1 \end{bmatrix} \begin{bmatrix} u_2 \\ u_3 \\ u_4 \\ u_5 \end{bmatrix} = \frac{\Delta X F_0}{E} \begin{bmatrix} 0 \\ 0 \\ 0 \\ 1 \end{bmatrix} ,$$

(2.96)

and the same result is obtained as given in Eq. (2.93).

(c) Starting from Eq. (2.56), the following FD scheme can be indicated for node i:

$$E \left(\frac{A_{i+1} - A_{i-1}}{2\Delta X} \times \frac{u_{i+1} - u_{i-1}}{2\Delta X} + A_i \times \frac{u_{i+1} - 2u_i + u_{i-1}}{\Delta X^2} \right) = 0 .$$

(2.97)

Writing the FD scheme in the following for nodes $i = 2, \ldots, 5$ and introducing by doing so a fictitious node at the right-hand boundary (see Fig. 2.12):

$$\text{node 2: } E \left(0 + \frac{2A}{\Delta X^2} (u_3 - 2u_2 + u_1) \right) = 0 ,$$

(2.98)

$$\text{node 3: } E \left(\frac{A - 2A}{2\Delta X} \times \frac{u_4 - u_2}{2\Delta X} + \frac{2A + A}{2\Delta X^2} \times (u_4 - 2u_3 + u_2) \right) = 0 ,$$

(2.99)

$$\text{node 4: } E \left(0 + \frac{A}{\Delta X^2} \times (u_5 - 2u_4 + u_3) \right) = 0 ,$$

(2.100)

node 5: $E\left(0 + \dfrac{A}{\Delta X^2} \times (u_6 - 2u_5 + u_4)\right) = 0.$ (2.101)

Replacing in Eq. (2.101) the fictitious node again by Eq. (2.90), i.e.,

$$u_6 = u_4 + \frac{2\Delta X F_0}{E_5 A_5},$$ (2.102)

the following system of equations can be stated:

$$\begin{bmatrix} -2 & 1 & 0 & 0 \\ 7 & -12 & 5 & 0 \\ 0 & 1 & -2 & 1 \\ 0 & 0 & 2 & -2 \end{bmatrix} \begin{bmatrix} u_2 \\ u_3 \\ u_4 \\ u_5 \end{bmatrix} = \frac{\Delta X F_0}{E A} \begin{bmatrix} 0 \\ 0 \\ 0 \\ -2 \end{bmatrix}.$$ (2.103)

The solution of this linear system of equations gives the unknown nodal values as:

$$\begin{bmatrix} u_2 \\ u_3 \\ u_4 \\ u_5 \end{bmatrix} = \frac{F_0 L}{E A} \begin{bmatrix} \frac{5}{28} \\ \frac{5}{14} \\ \frac{17}{28} \\ \frac{6}{7} \end{bmatrix} = \frac{F_0 L}{E A} \begin{bmatrix} 0.17857143 \\ 0.35714286 \\ 0.60714286 \\ 0.85714286 \end{bmatrix}.$$ (2.104)

2.5 Supplementary Problems

2.3 Finite difference approximation of a cantilevered rod with distributed load based on five domain nodes

Given is a cantilevered rod of length L with constant tensile stiffness $E A$ as shown in Fig. 2.13. The rod is loaded by a constant distributed load p_0. Use five domain nodes of equidistant spacing, i.e. $\Delta X = \frac{L}{4}$, for the finite difference approximation. Use only centered difference approximations of second order accuracy for the nodal evaluations and boundary conditions. Determine

- the horizontal displacement at the end of the rod, i.e. $X = L$,
- the analytical solution and

Fig. 2.13 Cantilevered rod loaded by a constant distributed load

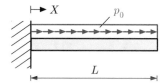

Fig. 2.14 Fixed-ended rod structure

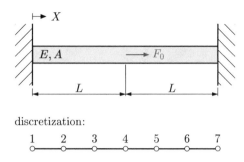

discretization:

• calculate the relative error between the analytical and finite difference solution.

2.4 Refined finite difference approximation of a cantilevered rod with distributed load

Reconsider Problem 2.3 and derive a general finite difference scheme for a larger number of nodes, i.e. $n > 5$.

2.5 Displacement distribution for a fixed-ended rod structure

Given is a rod structure as shown in Fig. 2.14. The structure is of length $2L$, cross-sectional area A, and Young's modulus E. The structure is fixed at both ends and loaded by a point load F_0 in the middle, i.e. at $X = L$. Calculate the displacement distribution based on seven grid nodes, i.e., $\Delta X = \frac{L}{3}$.

2.6 Displacement distribution for a fixed-ended stepped rod structure

Given is a rod structure as shown in Fig. 2.15. The structure is composed of two rods of different cross-sectional areas $A_\mathrm{I} = A$ and $A_\mathrm{II} = 2A$. Length L and Young's modulus E are the same for both rods. The structure is fixed at both ends and loaded by

Fig. 2.15 Rod structure fixed at both ends: **a** axial point load; **b** load per length

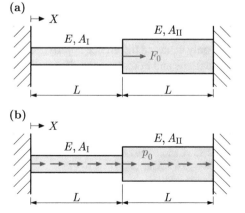

Fig. 2.16 Elongation of a
rod due to a distributed load
in the segment $a_1 \leq X \leq a_2$

Fig. 2.17 Bi-material rod
discretized by five grid
points

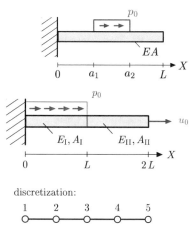

(a) a point load F_0 in the middle and
(b) a uniform distributed load p_0, i.e. a force per unit length.

Model the rod structure based on seven grid nodes and determine for both cases
the displacements.

2.7 Elongation of a rod due to distributed load

Calculate the elongation $u(X)$ of the rod shown in Fig. 2.16. The rod is loaded in the
section $a_1 = \frac{L}{3} \leq X \leq a_2 = \frac{2L}{3}$ by a constant distributed load p_0.
   Model the rod structure based on seven grid nodes, i.e., $\Delta X = \frac{L}{6}$, and compare
the finite difference solution to the exact analytical displacement distribution.

2.8 Elongation of a bi-material rod: finite difference solution and comparison with analytical solution

Given is a rod as shown in Fig. 2.17 which is made of two different sections with
axial stiffness $k_{\mathrm{I}} = E_{\mathrm{I}}A_{\mathrm{I}}$ and $k_{\mathrm{II}} = E_{\mathrm{II}}A_{\mathrm{II}}$. Each section is of length L and in the
left-hand section, i.e. $0 \leq X \leq L$, is a constant distributed load p_0 acting while the
right-hand end is elongated by u_0.
   Use five grid points to discretize the rod and calculate the displacements. Compare
the results with the analytical solution and sketch the distributions $u(X)$ for the case
$k_{\mathrm{I}} = 2k_{\mathrm{II}} = 1$, $L_{\mathrm{I}} = L_{\mathrm{II}} = 1$, $p_0 = 1$, $u_0 = 1$, and $E_{\mathrm{I}} = 2E_{\mathrm{II}} = 1$.

References

1. Altenbach H, Öchsner A (eds) (2020) Encyclopedia of continuum mechanics. Springer, Berlin
2. Bathe K-J (1996) Finite element procedures. Prentice-Hall, Upper Saddle River

3. Beer FP, Johnston ER Jr, DeWolf JT, Mazurek DF (2009) Mechanics of materials. McGraw-Hill, New York
4. Gere JM, Timoshenko SP (1991) Mechanics of materials. PWS-KENT Publishing Company, Boston
5. Hibbeler RC (2008) Mechanics of materials. Prentice Hall, Singapore
6. Öchsner A (2014) Elasto-plasticity of frame structure elements: modeling and simulation of rods and beams. Springer, Berlin
7. Öchsner A (2020) Partial differential equations of classical structural members: a consistent approach. Springer, Cham
8. Öchsner A (2020) Computational statics and dynamics: an introduction based on the finite element method. Springer, Singapore
9. Timoshenko S (1940) Strength of materials - part I elementary theory and problems. D. Van Nostrand Company, New York
10. Timoshenko SP, Goodier JN (1970) Theory of elasticity. McGraw-Hill, New York

Chapter 3
Investigation of Euler–Bernoulli Beams in the Elastic Range

3.1 The Basics of an Euler–Bernoulli Beam

A thin or Euler–Bernoulli beam is defined as a long prismatic body whose axial dimension is much larger than its transverse dimensions [2, 6, 8, 12, 13]. This structural member is only loaded perpendicular to its longitudinal body axis by forces (single forces F_Z or distributed loads q_Z) or moments (single moments M_Y or distributed moments m_Y). Perpendicular means that the line of application of a force or the direction of a moment vector forms a right angle with the X-axis, see Fig. 3.1. As a result of this loading, the deformation occurs only perpendicular to its main axis.

Derivations are restricted many times to the following simplifications:

- only applying to straight beams,
- no elongation along the X-axis,
- no torsion around the X-axis,
- deformations in a single plane (here: X-Z), i.e. symmetrical bending,
- infinitesimally small deformations and strains,
- simple cross sections, and
- the material is linear-elastic, i.e., constant Young's modulus E.

The three basic equations of continuum mechanics, i.e. the kinematics relationship, the constitutive law and the equilibrium equation, as well as their combination to the describing partial differential equation are summarized in Table 3.1.

Alternative formulations of the of the fourth order partial differential equations for a beam are collected in Table 3.2 where different types of loadings, geometry and bedding are differentiated. The last case in Table 3.2 refers to the elastic foundation of a beam which is also know in the literature as Winkler foundation [15]. The elastic foundation or Winkler foundation modulus k has in the case of beams[1] the unit of force per unit area.

Under the assumption of constant material ($E =$ const.) and geometric ($I_Y =$ const.) properties, the differential equation in Table 3.1 can be integrated four times

---

[1]In the general case, the unit of the elastic foundation modulus is force per unit area per unit length, i.e. $\frac{N}{m^2}/m = \frac{N}{m^3}$.

© The Author(s), under exclusive license to Springer Nature Switzerland AG 2021
A. Öchsner, *Structural Mechanics with a Pen*,
https://doi.org/10.1007/978-3-030-65892-2_3

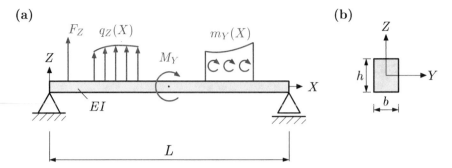

Fig. 3.1 General configuration for Euler–Bernoulli beam problems: **a** example of boundary conditions and external loads; **b** cross-sectional area (bending occurs in the X-Z plane)

for constant distributed load ($q_Z = q_0 = $ const.) to obtain the general analytical solution of the problem:

$$u_Z(X) = \frac{1}{E I_Y} \left(\frac{q_0 X^4}{24} + \frac{c_1 X^3}{6} + \frac{c_2 X^2}{2} + c_3 X + c_4 \right), \tag{3.1}$$

where the four constants of integration $c_i (i = 1, \ldots, 4)$ must be determined based on the boundary conditions (see Table 3.3). The following equations for the shear force $Q_Z(X)$, the bending moment $M_Y(X)$, and the rotation $\varphi_Y(X)$ were obtained based on one-, two- and three-times integration and might be useful to determine some of the constants of integration:

$$Q_Z(X) = -q_0 X - c_1, \tag{3.2}$$

$$M_Y(X) = -\frac{q_0 X^2}{2} - c_1 X - c_2, \tag{3.3}$$

$$\varphi_Y(X) = -\frac{\mathrm{d} u_Z(X)}{\mathrm{d}X} = -\frac{1}{E I_Y} \left(\frac{q_0 X^3}{6} + \frac{c_1 X^2}{2} + c_2 X + c_3 \right). \tag{3.4}$$

The internal reactions in a beam become visible if one cuts—at an arbitrary location X—the member in two parts. As a result, two opposite oriented shear forces Q_Z and bending moments M_Y can be indicated. Summing up the internal reactions from both parts must result in zero. Their positive direction is connected with the positive coordinate directions at the positive face (outward surface normal vector parallel to the positive X-axis). This means that at a positive face the positive reactions have the same direction as the positive coordinate axes, see Fig. 3.2.

Once the internal bending moment M_Y is known, the normal stress σ_X can be calculated:

$$\sigma_X(X, Z) = \frac{M_Y(X)}{I_Y} Z(X), \tag{3.5}$$

Table 3.1 Different formulations of the basic equations for a Bernoulli beam (bending occurs in the X-Z plane), with $\mathcal{L}_2(\ldots) = \frac{d^2(\ldots)}{dX^2}$

Specific formulation	General formulation [1]
Kinematics	
$\varepsilon_X(X, Z) = -Z\frac{d^2 u_Z(X)}{dX^2}$	$\varepsilon_X(X, Z) = -Z\mathcal{L}_2(u_Z(X))$
$\kappa = -\frac{d^2 u_Z(X)}{dX^2}$	$\kappa = -\mathcal{L}_2(u_Z(X))$
Constitution	
$\sigma_X(X, Z) = E\varepsilon_X(X, Z)$	$\sigma_X(X, Z) = C\varepsilon_X(X, Z)$
$M_Y(X) = EI_Y\kappa(X)$	$M_Y(X) = D\kappa(X)$
Equilibrium	
Force	
$\frac{dQ_Z(X)}{dX} = -q_Z(X)$	
Moment	
$\frac{dM_Y(X)}{dX} = Q_Z(X)$	
Combined	
$\frac{d^2 M_Y(X)}{dX^2} + q_Z(X) = 0$	$\mathcal{L}_2^T(M_Y(X)) + q_Z(X) = 0$
PDE	
$\frac{d^2}{dX^2}\left(EI_Y\frac{d^2 u_Z(X)}{dX^2}\right) - q_Z(X) = 0$	$\mathcal{L}_2^T(D\mathcal{L}_2(u_Z(X))) - q_Z(X) = 0$
$\frac{d}{dX}\left(EI_Y\frac{d^2 u_Z(X)}{dX^2}\right) = -Q_Z(X)$	
$EI_Y\frac{d^2 u_Z(X)}{dX^2} = -M_Y(X)$	

Fig. 3.2 Internal reactions for a continuum Euler–Bernoulli beam

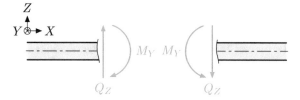

whereas the shear force Q_Z allows us to calculate the shear stress distribution. For a rectangular cross section (width b, height h, see Fig. 3.1) under the assumption that the shear stress is constant along the width, the following distribution is obtained [8]:

$$\tau_{XZ}(X, Z) = \frac{Q_Z(X)}{2I_Y}\left[\left(\frac{h}{2}\right)^2 - Z^2\right]. \tag{3.6}$$

Table 3.2 Different formulations of the partial differential equation for an Euler–Bernoulli beam in the X-Z plane (X-axis: right facing; Z-axis: upward facing)

Configuration	Partial differential equation
E, I_Y	$EI_Y \frac{\mathrm{d}^4 u_Z}{\mathrm{d}X^4} = 0$
$E(X)I_Y(X)$	$\frac{\mathrm{d}^2}{\mathrm{d}X^2}\left(E(X)I_Y(X)\frac{\mathrm{d}^2 u_Z}{\mathrm{d}X^2}\right) = 0$
$q_Z(X)$	$EI_Y \frac{\mathrm{d}^4 u_Z}{\mathrm{d}X^4} = q_Z(X)$
$m_Y(X)$	$EI_Y \frac{\mathrm{d}^4 u_Z}{\mathrm{d}X^4} = \frac{\mathrm{d}m_Y(X)}{\mathrm{d}X}$
$k(X)$	$EI_Y \frac{\mathrm{d}^4 u_Z}{\mathrm{d}X^4} = -k(X)u_Z$

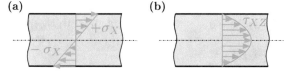

Fig. 3.3 Different stress distributions of an Euler–Bernoulli beam with rectangular cross section and linear-elastic material behavior: **a** normal stress and **b** shear stress (bending occurs in the X-Z plane)

Application of Hooke's law (i.e., $\sigma_X = E\varepsilon_X$ and $\tau_{XZ} = G\gamma_{XZ}$) allows us to calculate the normal and shear strains. Typical distributions of the two stress components in a beam element are shown in Fig. 3.3. It can be seen that normal stress distribution is linear while the shear stress distribution is parabolic over the cross section.

Finally, it should be noted here that the one-dimensional Euler–Bernoulli beam theory has its two-dimensional analogon in the form of Kirchhoff plates[2] [3–5, 7, 9, 14].

---

[2] Also called thin or shear-rigid plates.

Table 3.3 Different boundary conditions and corresponding reactions for a continuum Euler–Bernoulli beam (bending occurs in the X-Z plane)

Case	Boundary condition	Reaction
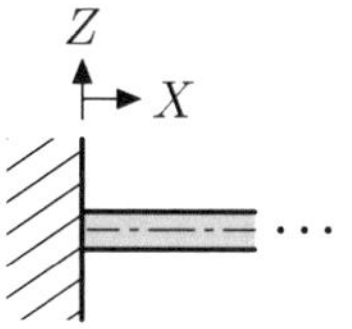	$u_Z(0) = 0,\ \varphi_Y(0) = 0$	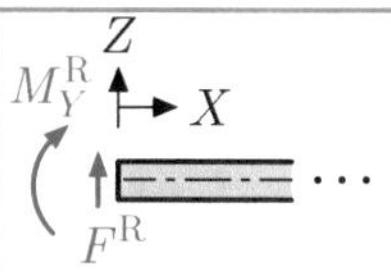
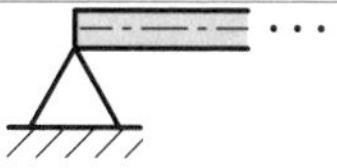	$u_Z(0) = 0,\ M_Y(0) = 0$	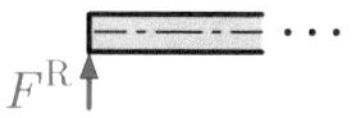
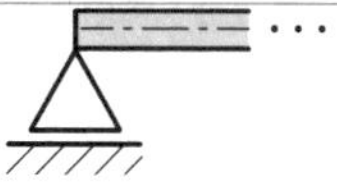	$u_Z(0) = 0,\ M_Y(0) = 0$	
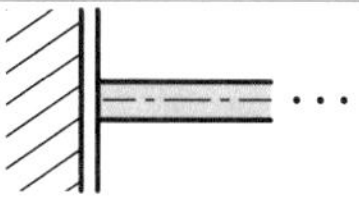	$\varphi_Y(0) = 0,\ Q_Z(0) = 0$	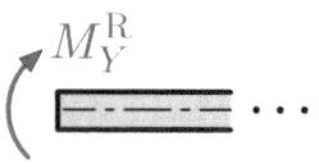
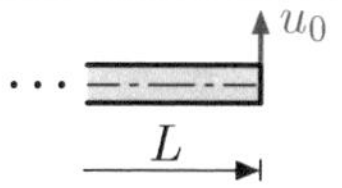	$u_Z(L) = u_0,\ M_Y(L) = 0$	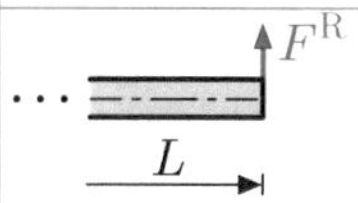
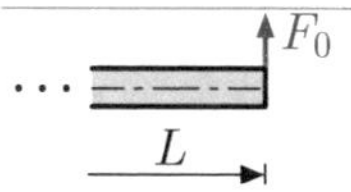	$Q_Z(L) = F_0,\ M_Y(L) = 0$	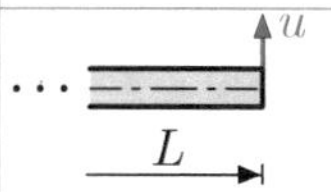
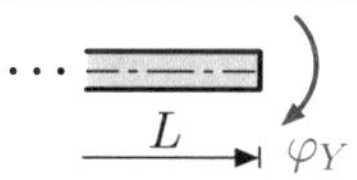	$\varphi_Y(L) = \varphi_0,\ Q_Z(L) = 0$	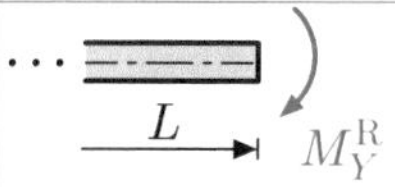
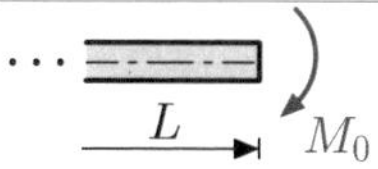	$M_Y(L) = M_0,\ Q_Z(L) = 0$	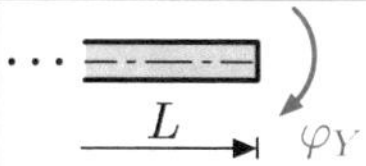
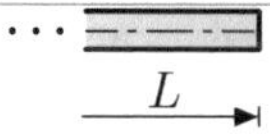	$M_Y(L) = 0,\ Q_Z(L) = 0$	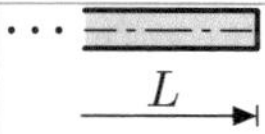

3.2   Constant Material and Geometry Parameters

Let us consider in the following a loaded Euler–Bernoulli beam of length L as shown in Fig. 3.4. The bending stiffness $E I_Y$, i.e. the material and geometric parameters, are considered to be constant in the following derivations. The boundary conditions are not shown in the figure and will be chosen in different ways throughout this chapter.

Fig. 3.4 Euler–Bernoulli
beam loaded with a
distributed load $q_Z(X)$

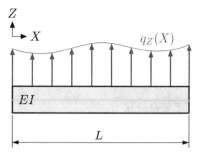

In the general case, a variable distributed load $q_Z(X)$ is acting along the length of
the beam.

The problem under consideration can be described according to Table 3.2 by the
following fourth-order differential equation

$$EI_Y \frac{\mathrm{d}^4 u_Z(X)}{\mathrm{d}X^4} = q_Z(X) \tag{3.7}$$

and appropriate boundary conditions. Taking from Table 1.1 the centered difference
scheme of second order accuracy, a finite difference approximation of the fourth-
order differential equation (3.7) can be written for node i as:

$$EI_Y \left(\frac{u_{i+2} - 4u_{i+1} + 6u_i - 4u_{i-1} + u_{i-2}}{\Delta X^4} \right) = q_i , \tag{3.8}$$

or as

$$\frac{EI_Y}{\Delta X^3} (u_{i+2} - 4u_{i+1} + 6u_i - 4u_{i-1} + u_{i-2}) = R_i , \tag{3.9}$$

where the equivalent nodal force R_i, resulting from a distributed load $q(X)$, is in
general given for an *inner* node i as: $R_i = \int_{-\Delta X/2}^{\Delta X/2} q(\hat{X})\mathrm{d}\hat{X}$. In this integral, the local
coordinate X has its origin at the location of node i. In the case of the boundary nodes,
the equivalent nodal loads must be calculated as $R_1 = \int_0^{\Delta X/2} q(\hat{X})\mathrm{d}\hat{X}$ (left-hand
node) or $R_n = \int_{-\Delta X/2}^0 q(\hat{X})\mathrm{d}\hat{X}$ (right-hand node) in order to completely distribute
the entire load $q(X)$ to the nodes. It should be noted here that—as in the case of
the finite element method—the finite difference method allows the action of forces
only at nodes. If a single force F_0 is acting at a node i, then this force should not be
considered on the right-hand side of Eq. (3.9).

The derivation of the second-order derivative was demonstrated in Chap. 1 and can
be used to calculate the centered difference scheme for the fourth-order derivative
as follows. The approximation is according to Table 1.1 given for node i as:

$$\left(\frac{\mathrm{d}^2 u}{\mathrm{d}X^2} \right)_i \approx \frac{u_{i+1} - 2u_i + u_{i-1}}{\Delta X^2} . \tag{3.10}$$

In a similar way, we can write the approximation for node $i + 1$ and $i - 1$ as

$$\left(\frac{d^2u}{dX^2}\right)_{i+1} \approx \frac{u_{i+2} - 2u_{i+1} + u_i}{\Delta X^2}, \tag{3.11}$$

$$\left(\frac{d^2u}{dX^2}\right)_{i-1} \approx \frac{u_i - 2u_{i-1} + u_{i-2}}{\Delta X^2}. \tag{3.12}$$

If we consider the left-hand sides (i.e., the fractions for the second-order derivatives) of the last three equations as functions u, we can introduce these three expressions in Eq. (3.10) to obtain the approximation for the second-order derivative of the second-order derivative, i.e., the fourth-order derivative, as:

$$\left(\frac{d^4u}{dX^4}\right)_i \approx \frac{u_{i+2} - 2u_{i+1} + u_i - 2(u_{i+1} - 2u_i + u_{i-1}) + u_i - 2u_{i-1} + u_{i-2}}{\Delta X^4}$$

$$= \frac{u_{i+2} - 4u_{i+1} + 6u_i - 4u_{i-1} + u_{i-2}}{\Delta X^4}. \tag{3.13}$$

3.3 Varying Material and Geometry Parameters

Let us consider in the following the case that the bending stiffness is a function of the Cartesian coordinate X. Thus, the generalized problem shown in Fig. 3.4 can be described in the domain $X \in [0, L]$ by the following partial differential equation

$$\frac{d^2}{dX^2}\left(E(X)I_Y(X)\frac{d^2u}{dX^2}\right) = q_Z(X), \tag{3.14}$$

or in the alternative formulations as

$$\frac{d}{dX}\left(E(X)I_Y(X)\frac{d^2u}{dX^2}\right) = -Q_Z(X), \tag{3.15}$$

$$E(X)I_Y(X)\frac{d^2u}{dX^2} = -M_Y(X). \tag{3.16}$$

Thus, we could start the consideration of varying parameters from three different formulations of the bending differential equation. Equation (3.15) has a similar structure as Eq. (2.33) and we will follow in our first attempt the line of reasoning presented at the beginning of Sect. 2.3 for a rod with varying parameters.

The product of the varying modulus and second moment of area can be combined in an auxiliary function

$$k(X) = E(X)I_Y(X),$$ (3.17)

and the differential equation (3.15) reads in a more general notation as:

$$\frac{\mathrm{d}}{\mathrm{d}X}\left(k(X)\frac{\mathrm{d}^2u}{\mathrm{d}X^2}\right) = -Q_Z(X).$$ (3.18)

Let us again introduce an auxiliary function $v(x)$ of the form:

$$v(X) = k(X)\frac{\mathrm{d}^2u}{\mathrm{d}X^2}.$$ (3.19)

Thus, the differential equation (3.18) can be expressed in a much more simpler way as:

$$\frac{\mathrm{d}v(X)}{\mathrm{d}X} = -Q_Z(X).$$ (3.20)

The first-order derivative $\frac{\mathrm{d}v}{\mathrm{d}X}$ is again approximated by the centered difference expression from Table 1.1. Replacing ΔX by $\frac{\Delta X}{2}$, we get

$$\left(\frac{\mathrm{d}v(X)}{\mathrm{d}X}\right)_i = \frac{v_{i+\frac{1}{2}} - v_{i-\frac{1}{2}}}{\Delta X},$$ (3.21)

where, for example, the notation '$i + \frac{1}{2}$' refers to the value of the function v in the middle of node i and $i + 1$, cf. Fig. 2.6 in Sect. 2.3. Based on the definition of the auxiliary function v in Eq. (3.19), the functional values of v in Eq. (3.21) at the intermediate locations can be stated based on centered difference approximations of second-order accuracy as ($\Delta X \rightarrow \Delta X/2$):

$$v_{i+\frac{1}{2}} = k_{i+\frac{1}{2}}\left(\frac{\mathrm{d}^2u}{\mathrm{d}X^2}\right)_{i+\frac{1}{2}} = k_{i+\frac{1}{2}}\frac{u_{i+1} - 2u_{i+\frac{1}{2}} + u_i}{\Delta X^2/4},$$ (3.22)

$$v_{i-\frac{1}{2}} = k_{i-\frac{1}{2}}\left(\frac{\mathrm{d}^2u}{\mathrm{d}X^2}\right)_{i-\frac{1}{2}} = k_{i-\frac{1}{2}}\frac{u_i - 2u_{i-\frac{1}{2}} + u_{i-1}}{\Delta X^2/4}.$$ (3.23)

However, this approach would introduce displacements at intermediate nodes, i.e., $u_{i+\frac{1}{2}}$ and $u_{i-\frac{1}{2}}$, and we will not proceed with this derivation.

Alternatively, we apply[3] the product rule of differential calculus to Eq. (3.15). Assuming constant material properties, this gives:

---

[3] Similar to Eq. (2.55) in Sect. 2.3.

$$E \left(\frac{dI_Y(X)}{dX} \frac{d^2u}{dX^2} + I_Y(X)\frac{d^3u}{dX^3} \right) = -Q_Z(X). \tag{3.24}$$

Introduction of centered difference schemes for the second- and third-order derivative according to Table 1.1 gives:

$$E \left(\left(\frac{dI_Y(X)}{dX} \right)_i \frac{u_{i+1} - 2u_i + u_{i-1}}{\Delta X^2} + I_{Y,i} \frac{u_{i+2} - 2u_{i+1} + 2u_{i-1} - u_{i-2}}{2\Delta X^3} \right) = -Q_{Z,i}, \tag{3.25}$$

where the gradient of the second moment of area function can be approximated based on a centered difference scheme as

$$\frac{dI_Y(X)}{dX} = \frac{I_{Y,i+1} - I_{Y,i-1}}{2\Delta X}. \tag{3.26}$$

Another approach can be based on the moment formulation of the partial differential equation according to (3.16) to give for node i:

$$E_i I_{Y,i} \frac{d^2u}{dX^2} \bigg|_i = E_i I_{Y,i} \frac{u_{i+1} - 2u_i + u_{i-1}}{\Delta X^2} = -M_{Y,i}. \tag{3.27}$$

3.4 Solved Problems

3.1 Example: Finite difference approximation of a simply supported and cantilevered beam loaded by a single force

Given is an Euler–Bernoulli beam with different supports as shown in Fig. 3.5. The bending stiffness EI_Y is constant and the length is equal to L. The simply supported beam (a) is loaded in the middle by a single force F_y while the cantilevered beam (b) is loaded at its right-hand end by a single force F_0. Derive for both cases a finite difference approximation based on five grid points, i.e. an equidistant spacing of $\Delta X = \frac{L}{4}$.

Determine for both cases

- the displacement of the beam at the force application point,
- the analytical solution and
- calculate the relative error between the analytical and finite difference solution.

3.1 Solution

(a) The finite difference discretization of the simply supported beam is shown in Fig. 3.6 where at first only the inner nodes will be considered.

It should be noted here that the fourth-order differential equation according to Eq. (3.9) does not allow to account for external single forces (in our case: F_0).

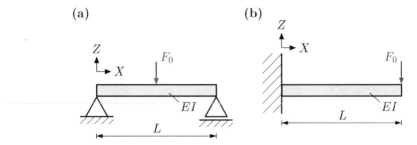

Fig. 3.5 a Simply supported and **b** cantilevered Euler–Bernoulli beam loaded by a single force

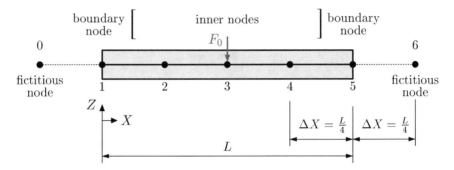

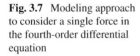

Fig. 3.6 Finite difference discretization of the simply supported Euler–Bernoulli beam

Fig. 3.7 Modeling approach
to consider a single force in
the fourth-order differential
equation

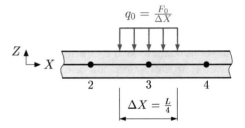

The equivalent nodal force R_i is obtained from distributed loads q_X and should not be confused with external single loads. If the derivations should be based on the fourth-order differential equation, then a modeling approach as shown in Fig. 3.7 can be applied. Thus, we understand the single force F_0 as the integral value of a distributed load q_0. Obviously, this is no more exactly the same load case as shown in Fig. 3.5a. Nevertheless, it allows us to proceed the derivations based on the fourth-order differential equation.

Evaluation of the finite difference approximation of the fourth-order differential equation according to Eq. (3.9) at the inner nodes $i = 2, \ldots, 4$ gives:

node 2: $\dfrac{EI_Y}{\Delta X^3}(u_4 - 4u_3 + 6u_2 - 4u_1 + u_0) = 0$,   (3.28)

node 3: $\dfrac{EI_Y}{\Delta X^3}(u_5 - 4u_4 + 6u_3 - 4u_2 + u_1) = -F_0 (= q_0\Delta X)$,   (3.29)

node 4: $\dfrac{EI_Y}{\Delta X^3}(u_6 - 4u_5 + 6u_4 - 4u_3 + u_2) = 0$.   (3.30)

The vertical displacement is zero at both ends and it can be immediately concluded that $u_1 = u_5 = 0$. The fictitious nodes $i = 0$ and $i = 6$ outside the domain can be eliminated based on the boundary condition that the moment must be equal to zero at the supports, i.e. $M_Y(X = 0) = M_Y(X = L) = 0$. Application of the bending differential equation in the form with the bending moment according to Table 3.1, i.e. $EI_Y\frac{d^2u}{dX^2} = -M_Y$, and the centered finite difference approximation of the second order derivative according to Table 1.1, the following two conditions can be derived:

$$EI_Y \left.\frac{d^2u}{dX^2}\right|_1 = EI_Y\frac{u_2 - 2u_1 + u_0}{\Delta X^2} = 0,$$ (3.31)

$$EI_Y \left.\frac{d^2u}{dX^2}\right|_5 = EI_Y\frac{u_6 - 2u_5 + u_4}{\Delta X^2} = 0,$$ (3.32)

from which the two conditions $u_0 = -u_2$ and $u_6 = -u_4$ can be obtained. Introducing these relationships for the fictitious nodes and the condition of zero displacement at the supports into the system of equations according to (3.28)–(3.30) gives:

node 2: $\dfrac{EI_Y}{\Delta X^3}(5u_2 - 4u_3 + u_4) = 0$,   (3.33)

node 3: $\dfrac{EI_Y}{\Delta X^3}(-4u_2 + 6u_3 - 4u_4) = -F_0$,   (3.34)

node 4: $\dfrac{EI_Y}{\Delta X^3}(u_2 - 4u_3 + 5u_4) = 0$,   (3.35)

or in matrix notation:

$$\frac{EI_Y}{\Delta X^3}\begin{bmatrix} 5 & -4 & 1 \\ -4 & 6 & -4 \\ 1 & -4 & 5 \end{bmatrix}\begin{bmatrix} u_2 \\ u_3 \\ u_4 \end{bmatrix} = \begin{bmatrix} 0 \\ -F_0 \\ 0 \end{bmatrix}.$$ (3.36)

The solution of this linear system of equations gives the unknown nodal values as:

$$u_2 = -\frac{F_0\Delta X^3}{EI_Y}, \quad u_3 = -\frac{3F_0\Delta X^3}{4EI_Y}, \quad u_4 = -\frac{F_0\Delta X^3}{EI_Y},$$ (3.37)

or with $\Delta X = \frac{L}{4}$ as:

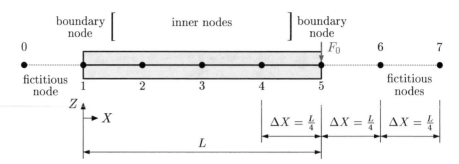

Fig. 3.8 Finite difference discretization of the cantilevered Euler–Bernoulli beam

Fig. 3.9 Modeling approach to consider a single force in the fourth-order differential equation

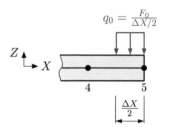

$$u_2 = -\frac{F_0 L^3}{64 E I_Y} \; , \; u_3 = -\frac{3 F_0 L^3}{128 E I_Y} \; , \; u_4 = -\frac{F_0 L^3}{64 E I_Y} \; . \tag{3.38}$$

The analytical solution for the displacement at the force application point can be taken from [10] as $\frac{-F_0 L^3}{48 E I_Y}$ and the relative error is obtained as:

$$\text{relative error} = \frac{\frac{3}{128} - \frac{1}{48}}{\frac{1}{48}} \times 100 = 12.5\% \, . \tag{3.39}$$

(b) The finite difference discretization of the cantilevered beam is shown in Fig. 3.8.

It should be noted here that the fourth-order differential equation according to Eq. (3.9) does not allow to account for external single forces (in our case: F_0 at the right-hand boundary). The equivalent nodal force R_i is obtained from distributed loads q_X and should not be confused with external single loads. If the derivations should be based on the fourth-order differential equation, then a modeling approach as shown in Fig. 3.9 can be applied. Thus, we understand the single force F_0 as the integral value of a distributed load q_0, which is acting over a length of $\Delta X/2$. Obviously, this is no more exactly the same load case as shown in Fig. 3.5b, i.e., a point load. Nevertheless, it allows us to proceed the derivations based on the fourth-order differential equation.

Evaluation of the finite difference approximation of the inner nodes gives similar to part (a) the following three equations:

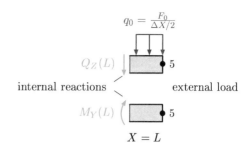

Fig. 3.10 Equilibrium between internal reactions and external load at the right-hand boundary ($X = L$) based on the modeling approach provided in Fig. 3.9

$$\text{node 2: } \frac{EI_Y}{\Delta X^3} (u_4 - 4u_3 + 6u_2 - 4u_1 + u_0) = 0, \qquad (3.40)$$

$$\text{node 3: } \frac{EI_Y}{\Delta X^3} (u_5 - 4u_4 + 6u_3 - 4u_2 + u_1) = 0, \qquad (3.41)$$

$$\text{node 4: } \frac{EI_Y}{\Delta X^3} (u_6 - 4u_5 + 6u_4 - 4u_3 + u_2) = 0. \qquad (3.42)$$

The vertical displacement is zero at the left-hand boundary because of the fixed support and it follows immediately that $u_1 = 0$ holds. In addition, the rotation is zero at the fixed support, i.e. $\frac{du}{dX}\big|_1 = 0$, and a centered finite difference approach according to Table 1.1 gives the condition $u_0 = u_2$.

The equilibrium between the internal reactions and the external load (according to our modeling approach provided in Fig. 3.9), cf. Fig. 3.10, and the application of the bending differential equations in the second and third form of Table 3.1 gives the following conditions at the right-hand boundary:

$$EI_Y \left.\frac{d^2u}{dX^2}\right|_5 = -M_Y(X = L) = 0, \qquad (3.43)$$

$$EI_Y \left.\frac{d^3u}{dX^3}\right|_5 = -Q_Z(X = L) = 0. \qquad (3.44)$$

Application of the centered difference approximations of the derivatives as given in Table 1.1, gives

$$EI_Y \times \frac{u_6 - 2u_5 + u_4}{\Delta X^2} = 0 \quad \text{or} \quad u_6 = 2u_5 - u_4. \qquad (3.45)$$

The evaluation of the condition for the shear force gives in the case of the centered difference scheme

$$EI_Y \times \frac{u_7 - 2u_6 + 2u_4 - u_3}{2\Delta X^3} = 0, \qquad (3.46)$$

which introduces as $i = 7$ a second fictitious node at the right-hand boundary. To overcome this problem, the finite difference approximation for the boundary node $i = 5$ can be written as:

$$\text{node 5: } \frac{E I_Y}{\Delta X^3} (u_7 - 4u_6 + 6u_5 - 4u_4 + u_3) = -F_0 . \tag{3.47}$$

Introduction of the relationships for u_0, u_1, u_6 and u_7 which are obtained from the boundary conditions in the finite difference approximations for nodes 2 till 5 gives the following system of equations:

$$\text{node 2: } \frac{E I_Y}{\Delta X^3} (7u_2 - 4u_3 + u_4) = 0 , \tag{3.48}$$

$$\text{node 3: } \frac{E I_Y}{\Delta X^3} (-4u_2 + 6u_3 - 4u_4 + u_5) = 0 , \tag{3.49}$$

$$\text{node 4: } \frac{E I_Y}{\Delta X^3} (u_2 - 4u_3 + 5u_4 - 2u_5) = 0 , \tag{3.50}$$

$$\text{node 5: } \frac{E I_Y}{\Delta X^3} (2u_3 - 4u_4 + 2u_5) = -F_0 , \tag{3.51}$$

or in matrix form:

$$\frac{E I_Y}{\Delta X^3} \begin{bmatrix} 7 & -4 & 1 & 0 \\ -4 & 6 & -4 & 1 \\ 1 & -4 & 5 & -2 \\ 0 & 2 & -4 & 2 \end{bmatrix} \begin{bmatrix} u_2 \\ u_3 \\ u_4 \\ u_5 \end{bmatrix} = \begin{bmatrix} 0 \\ 0 \\ 0 \\ -F_0 \end{bmatrix} . \tag{3.52}$$

The solution of this linear system of equations gives the unknown nodal values as:

$$u_2 = -\frac{F_0 \Delta X^3}{E I_Y} , \quad u_3 = -\frac{7 F_0 \Delta X^3}{2 E I_Y} , \quad u_4 = -\frac{7 F_0 \Delta X^3}{E I_Y} , \quad u_5 = -\frac{11 F_0 \Delta X^3}{E I_Y} , \tag{3.53}$$

or with $\Delta X = \frac{L}{4}$ as:

$$u_2 = -\frac{F_0 L^3}{64 E I_Y} , \quad u_3 = -\frac{7 F_0 L^3}{128 E I_Y} , \quad u_4 = -\frac{7 F_0 L^3}{64 E I_Y} , \quad u_5 = -\frac{11 F_0 L^3}{64 E I_Y} . \tag{3.54}$$

The analytical solution for the displacement at the force application point can be taken from [10] as $\frac{-F_0 L^3}{3 E I_Y}$ and the relative error is obtained as:

$$\text{relative error} = \left| \frac{\frac{11}{64} - \frac{1}{3}}{\frac{1}{3}} \right| \times 100 = 48.438\% . \tag{3.55}$$

A different way of solution can be chosen by avoiding the second ($i = 7$) fictitious node at the right-hand end. To this end, a backward finite difference approximation (cf. Table 1.1) can be introduced into the condition for the internal shear force[4] at the right-hand end:

---

[4]The value of the internal shear force is now determined based on the real load condition, i.e., Fig. 3.5b, and not based on the modeling approach provided in Fig. 3.9. Thus, we obtain: $Q_Z(X = L) = -F_0$.

$$EI_Y \underbrace{\frac{5u_5 - 18u_4 + 24u_3 - 14u_2 + 3u_1}{2\Delta X^3}}_{\frac{d^3u}{dX^3}\big|_5} = -Q_Z(X = L) = F_0 . \qquad (3.56)$$

The fifth equation reads now under consideration of the boundary condition at the left-hand end ($u_1 = 0$)

$$\text{node 5: } 5u_5 - 18u_4 + 24u_3 - 14u_2 = \frac{2\Delta X^3 F_0}{EI_Y} , \qquad (3.57)$$

and the final system of equations is obtained as:

$$\frac{EI_Y}{\Delta X^3} \begin{bmatrix} 7 & -4 & 1 & 0 \\ -4 & 6 & -4 & 1 \\ 1 & -4 & 5 & -2 \\ -14 & 24 & -18 & 5 \end{bmatrix} \begin{bmatrix} u_2 \\ u_3 \\ u_4 \\ u_5 \end{bmatrix} = \begin{bmatrix} 0 \\ 0 \\ 0 \\ 2F_0 \end{bmatrix} . \qquad (3.58)$$

The solution of this linear system of equations gives with $\Delta X = \frac{L}{4}$ the unknown nodal values as:

$$u_2 = -\frac{F_0 L^3}{32EI_Y} , \quad u_3 = -\frac{7F_0 L^3}{64EI_Y} , \quad u_4 = -\frac{7F_0 L^3}{32EI_Y} , \quad u_5 = -\frac{11F_0 L^3}{32EI_Y} , \qquad (3.59)$$

and the relative error is obtained as:

$$\text{relative error} = \frac{\frac{11}{32} - \frac{1}{3}}{\frac{1}{3}} \times 100 = 3.126\% . \qquad (3.60)$$

3.2 Example: Finite difference approximation of a simply supported and cantilevered beam loaded by a distributed load

Given is an Euler–Bernoulli beam with different supports as shown in Fig. 3.11. The bending stiffness EI_Y is constant and the length is equal to L. The simply supported beam (a) and the cantilevered beam (b) are loaded by a constant distributed load q_0. Derive for both cases a finite difference approximation based on five grid points, i.e. an equidistant spacing of $\Delta X = \frac{L}{4}$.

Determine for both cases

- the displacement in the middle of the beam for (a) and at the free tip for case (b),
- the analytical solution and
- calculate the relative error between the analytical and finite difference solution.

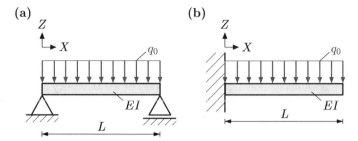

Fig. 3.11 a Simply supported and **b** cantilevered Euler–Bernoulli beam loaded by a distributed load

3.2 Solution

The problem of this example is very similar to the previous example Problem 3.1 and major parts can be handled in a similar way.

(a) For the case of the simply supported beam we write—as in the case of example Problem 3.1—the finite difference approximation according to Eq. (3.9) at the inner nodes $i = 2, \ldots, 4$ under consideration of the distributed load as:

$$\text{node 2:} \quad \frac{E I_Y}{\Delta X^3} (u_4 - 4u_3 + 6u_2 - 4u_1 + u_0) = -q_0 \Delta X, \qquad (3.61)$$

$$\text{node 3:} \quad \frac{E I_Y}{\Delta X^3} (u_5 - 4u_4 + 6u_3 - 4u_2 + u_1) = -q_0 \Delta X, \qquad (3.62)$$

$$\text{node 4:} \quad \frac{E I_Y}{\Delta X^3} (u_6 - 4u_5 + 6u_4 - 4u_3 + u_2) = -q_0 \Delta X. \qquad (3.63)$$

Consideration of the boundary conditions, i.e. $u_1 = u_5 = 0$, $u_0 = -u_2$ and $u_6 = -u_4$, results in the following system of equations:

$$\frac{E I_Y}{\Delta X^3} \begin{bmatrix} 5 & -4 & 1 \\ -4 & 6 & -4 \\ 1 & -4 & 5 \end{bmatrix} \begin{bmatrix} u_2 \\ u_3 \\ u_4 \end{bmatrix} = \begin{bmatrix} -q_0 \Delta X \\ -q_0 \Delta X \\ -q_0 \Delta X \end{bmatrix}. \qquad (3.64)$$

The solution of this linear system of equations gives the unknown nodal values as:

$$u_2 = -\frac{5 q_0 \Delta X^4}{2 E I_Y}, \quad u_3 = -\frac{7 q_0 \Delta X^4}{2 E I_Y}, \quad u_4 = -\frac{5 q_0 \Delta X^4}{2 E I_Y}, \qquad (3.65)$$

or with $\Delta X = \frac{L}{4}$ as:

$$u_2 = -\frac{5 q_0 L^4}{512 E I_Y}, \quad u_3 = -\frac{7 q_0 L^4}{512 E I_Y}, \quad u_4 = -\frac{5 q_0 L^4}{512 E I_Y}. \qquad (3.66)$$

Fig. 3.12 Equivalent force system for the distributed load problem shown in Fig. 3.11b:
$R_1 = R_5 = -\frac{q_0 \Delta X}{2}$,
$R_2 = R_3 = R_4 = q_0 \Delta X$

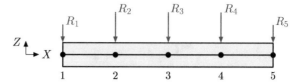

The analytical solution for the displacement in the middle of the beam can be taken from [10] as $\frac{-5 q_0 L^4}{384 E I_Y}$ and the relative error is obtained as:

$$\text{relative error} = \frac{\frac{7}{512} - \frac{5}{384}}{\frac{5}{384}} \times 100 = 5.0\% . \tag{3.67}$$

The graphical comparison between this finite difference approach and the analytical solution is presented in Fig. 3.13a where it can be seen that the finite difference approach slightly overestimates the deformation between the boundary supports.

(b) As in the previous example Problem 3.1, we write the finite difference approximation for nodes $i = 2, \ldots, 5$ under consideration of the distributed load as[5]:

$$\text{node 2:} \quad \frac{E I_Y}{\Delta X^3} (u_4 - 4u_3 + 6u_2 - 4u_1 + u_0) = -q_0 \Delta X , \tag{3.68}$$

$$\text{node 3:} \quad \frac{E I_Y}{\Delta X^3} (u_5 - 4u_4 + 6u_3 - 4u_2 + u_1) = -q_0 \Delta X , \tag{3.69}$$

$$\text{node 4:} \quad \frac{E I_Y}{\Delta X^3} (u_6 - 4u_5 + 6u_4 - 4u_3 + u_2) = -q_0 \Delta X , \tag{3.70}$$

$$\text{node 5:} \quad \frac{E I_Y}{\Delta X^3} (u_7 - 4u_6 + 6u_5 - 4u_4 + u_3) = -q_Y \frac{\Delta X}{2} . \tag{3.71}$$

Consideration of the boundary conditions at node 1 gives as in the previous example $u_1 = 0$ and $u_0 = u_2$.

The equilibrium between the internal reactions and the external load based on the equivalent force system (cf. Fig. 3.12) gives here

$$E I_Y \frac{d^2 u}{d X^2}\bigg|_5 = -M_Y(X = L) = 0 , \tag{3.72}$$

$$E I_Y \frac{d^3 u}{d X^3}\bigg|_5 = -Q_Z(X = L) = q_0 \frac{\Delta X}{2} , \tag{3.73}$$

from which the following two conditions can be derived if a centered difference approximation is introduced:

---

[5]It is important to consider for the equivalent nodal force at the boundary node 5 only the effective length of $\frac{\Delta X}{2}$: $R_5 = q_0 \frac{\Delta X}{2}$.

$$u_6 = 2u_5 - u_4 ,$$ (3.74)

$$u_7 = 4u_5 - 4u_4 + u_3 + \frac{q_0 \Delta X^4}{E I_Y} .$$ (3.75)

Thus, the final system of equations is given as:

$$\frac{E I_Y}{\Delta X^3} \begin{bmatrix} 7 & -4 & 1 & 0 \\ -4 & 6 & -4 & 1 \\ 1 & -4 & 5 & -2 \\ 0 & 2 & -4 & 2 \end{bmatrix} \begin{bmatrix} u_2 \\ u_3 \\ u_4 \\ u_5 \end{bmatrix} = \begin{bmatrix} -q_0 \Delta X \\ -q_0 \Delta X \\ -q_0 \Delta X \\ -q_0 \frac{3\Delta X}{2} \end{bmatrix} .$$ (3.76)

The solution of this linear system of equations gives the unknown nodal values as:

$$u_2 = \frac{9 q_0 \Delta X^4}{2 E I_Y} , \quad u_3 = -\frac{57 q_0 \Delta X^4}{4 E I_Y} , \quad u_4 = -\frac{53 q_0 \Delta X^4}{2 E I_Y} , \quad u_5 = -\frac{79 q_0 \Delta X^4}{2 E I_Y} ,$$ (3.77)

or with $\Delta X = \frac{L}{4}$ as:

$$u_2 = -\frac{9 q_0 L^4}{512 E I_Y} , \quad u_3 = -\frac{57 q_0 L^4}{1024 E I_Y} , \quad u_4 = -\frac{53 q_0 L^4}{512 E I_Y} , \quad u_5 = -\frac{79 q_0 L^4}{512 E I_Y} .$$ (3.78)

The analytical solution for the displacement at the beam tip can be taken from [10] as $\frac{-q_0 L^4}{8 E I_Y}$ and the relative error is obtained as:

$$\text{relative error} = \frac{\frac{79}{512} - \frac{1}{8}}{\frac{1}{8}} \times 100 = 23.438\% .$$ (3.79)

A different approach to the solution can be chosen by replacing the boundary condition (3.73) by the analytical boundary condition which is not based on the equivalent system, i.e.

$$E I_Y \left. \frac{d^3 u}{dX^3} \right|_5 = -Q_Z(X = L) = 0 ,$$ (3.80)

and the condition for the fictitious node 7 is obtained as

$$u_7 = 4u_5 - 4u_4 + u_3 .$$ (3.81)

Thus, the final system of equations is obtained for this case as:

$$\frac{E I_Y}{\Delta X^3} \begin{bmatrix} 7 & -4 & 1 & 0 \\ -4 & 6 & -4 & 1 \\ 1 & -4 & 5 & -2 \\ 0 & 2 & -4 & 2 \end{bmatrix} \begin{bmatrix} u_2 \\ u_3 \\ u_4 \\ u_5 \end{bmatrix} = \begin{bmatrix} -q_0 \Delta X \\ -q_0 \Delta X \\ -q_0 \Delta X \\ -q_0 \frac{\Delta X}{2} \end{bmatrix} .$$ (3.82)

The solution of this linear system of equations gives with $\Delta X = \frac{L}{4}$ the unknown nodal values as:

$$u_2 = -\frac{7q_0 L^4}{512EI_Y} \ , \ u_3 = -\frac{43q_0 L^4}{1024EI_Y} \ , \ u_4 = -\frac{39q_0 L^4}{512EI_Y} \ , \ u_5 = -\frac{57q_0 L^4}{512EI_Y} \ ,$$
(3.83)

and a much smaller relative error is obtained:

$$\text{relative error} = \frac{\frac{57}{512} - \frac{1}{8}}{\frac{1}{8}} \times 100 = 10.938\% \ .$$
(3.84)

The graphical comparison between these two approaches and the analytical solution is given in Fig. 3.13. Depending on the approach, a different behavior, i.e. to over- or underestimate the analytical solution is obtained.

3.3 Example: Finite difference approximation of a simply supported beam loaded by a varying distributed load

Given is a simply supported Euler–Bernoulli beam as shown in Fig. 3.14. The bending stiffness EI_Y is constant and the length is equal to L. The simply supported beam is loaded by a linearly varying distributed load $q_Z(X)$. Derive a finite difference approximation based on five grid points, i.e. an equidistant spacing of $\Delta X = \frac{L}{4}$. Use centered difference schemes where the truncation error is of order ΔX^2.

Determine

- the displacements at the grid points,
- the analytical solution and
- calculate the relative error between the analytical and finite difference solution for the displacement in the middle of the beam.

3.3 Solution

The function of the distributed load can be expressed as

$$q_Z(X) = -q_0 \left(1 + \frac{X}{L} \right) \ ,$$
(3.85)

whereas the values at the five grid points are collected in Table 3.4.

The finite difference discretization of the simply supported beam is shown in Fig. 3.15 where at first only the inner nodes will be considered.

Evaluation of the finite difference approximation of the fourth-order differential equation according to Eq. (3.9) at the inner nodes $i = 2, \ldots, 4$ gives:

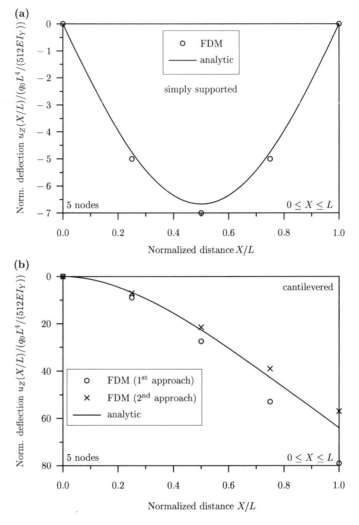

Fig. 3.13 Displacement along the normalized beam axis obtained from a FDM approach compared to the analytical solution: **a** simply supported and **b** cantilevered beam, cf. Fig. 3.11

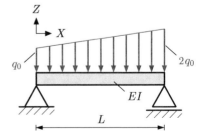

Fig. 3.14 Simply supported Euler–Bernoulli beam loaded by a varying distributed load

Table 3.4 Values of the distributed load at the grid points (see Fig. 3.15)

Grid point	Coordinate X	$q_Z(X)$
1	0	$-q_0$
2	$\frac{L}{4}$	$-\frac{5}{4} q_0$
3	$\frac{L}{2}$	$-\frac{3}{2} q_0$
4	$\frac{3L}{4}$	$-\frac{7}{4} q_0$
5	L	$-2q_0$

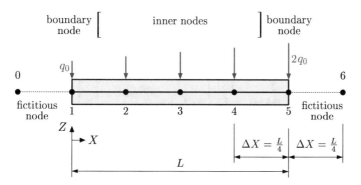

Fig. 3.15 Finite difference discretization of the cantilevered Euler–Bernoulli beam shown in Fig. 3.14

$$\text{node 2:} \quad \frac{E I_Y}{\Delta X^3} (u_4 - 4u_3 + 6u_2 - 4u_1 + u_0) = -\Delta X \frac{5q_0}{4}, \quad (3.86)$$

$$\text{node 3:} \quad \frac{E I_Y}{\Delta X^3} (u_5 - 4u_4 + 6u_3 - 4u_2 + u_1) = -\Delta X \frac{3q_0}{2}, \quad (3.87)$$

$$\text{node 4:} \quad \frac{E I_Y}{\Delta X^3} (u_6 - 4u_5 + 6u_4 - 4u_3 + u_2) = -\Delta X \frac{7q_0}{4}. \quad (3.88)$$

The vertical displacement is zero at both ends and it can be immediately concluded that $u_1 = u_5 = 0$. The fictitious nodes $i = 0$ and $i = 6$ outside the domain can be eliminated based on the boundary condition that the moment must be equal to zero at the supports, i.e. $M_Y(X = 0) = M_Y(X = L) = 0$. Application of the bending differential equation in the form with the bending moment according to Table 3.1, i.e. $E I_Y \frac{d^2 u}{dX^2} = -M_Y$, and the centered finite difference approximation of the second-order derivative according to Table 1.1, the following two conditions can be derived:

$$E I_Y \left. \frac{d^2 u}{dX^2} \right|_1 = E I_Y \frac{u_2 - 2u_1 + u_0}{\Delta X^2} = 0, \quad (3.89)$$

$$E I_Y \left. \frac{d^2 u}{dX^2} \right|_5 = E I_Y \frac{u_6 - 2u_5 + u_4}{\Delta X^2} = 0, \quad (3.90)$$

from which the two conditions $u_0 = -u_2$ and $u_6 = -u_4$ can be obtained. Introducing these relationships for the fictitious nodes and the condition of zero displacement at the supports into the system of equations according to (3.86)–(3.88) gives:

$$\text{node 2: } \frac{EI_Y}{\Delta X^3}(5u_2 - 4u_3 + u_4) = -\Delta X \frac{5q_0}{4}, \tag{3.91}$$

$$\text{node 3: } \frac{EI_Y}{\Delta X^3}(-4u_2 + 6u_3 - 4u_4) = -\Delta X \frac{3q_0}{2}, \tag{3.92}$$

$$\text{node 4: } \frac{EI_Y}{\Delta X^3}(u_2 - 4u_3 + 5u_4) = -\Delta X \frac{7q_0}{4}, \tag{3.93}$$

or in matrix notation with $\Delta X = L/4$:

$$\begin{bmatrix} 5 & -4 & 1 \\ -4 & 6 & -4 \\ 1 & -4 & 5 \end{bmatrix} \begin{bmatrix} u_2 \\ u_3 \\ u_4 \end{bmatrix} = -\frac{q_0 L^4}{1024 EI_Y} \begin{bmatrix} 5 \\ 6 \\ 7 \end{bmatrix}. \tag{3.94}$$

The solution of this linear system of equations gives the unknown nodal values as:

$$\begin{bmatrix} u_2 \\ u_3 \\ u_4 \end{bmatrix} = -\frac{q_0 L^4}{4096 EI_Y} \begin{bmatrix} 59 \\ 84 \\ 61 \end{bmatrix} \approx -\frac{q_0 L^4}{EI_Y} \begin{bmatrix} 0.014404 \\ 0.020508 \\ 0.014893 \end{bmatrix}. \tag{3.95}$$

The analytical solution for the displacement at the force application point can be derived from the differential equation through four times integration ($c_1 = 2q_0 L/3$, $c_2 = 0$, $c_3 = -11q_0 L^3/180$, $c_4 = 0$) as $-0.0195\frac{q_0 L^4}{EI_Y}$ and the relative error is obtained as:

$$\text{relative error} = \frac{0.020508 - 0.0195}{0.0195} \times 100 = 5.2\%. \tag{3.96}$$

3.4 Example: Refined finite difference approximation of a simply supported and cantilevered beam loaded by a single force

Given is an Euler–Bernoulli beam with different supports as shown in Fig. 3.16. The bending stiffness EI_Y is constant and the length is equal to L. The simply supported beam (a) is loaded in the middle by a single force while the cantilevered beam (b) is loaded at its right-hand end by a single force F_0. Derive for both cases a finite difference approximation based on seven grid points, i.e. an equidistant spacing of $\Delta X = \frac{L}{6}$.

Determine for both cases

- the displacement of the beam at the force application point,
- the analytical solution and
- calculate the relative error between the analytical and finite difference solution.

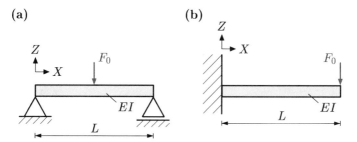

Fig. 3.16 Refined analysis: **a** Simply supported and **b** cantilevered Euler–Bernoulli beam loaded by a single force

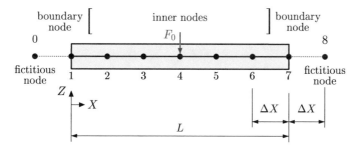

Fig. 3.17 Finite difference discretization of the simply supported Euler–Bernoulli beam based on seven grid nodes

Start from the fourth-order as well as the second-order differential equation and compare the results. Derive from the results a general scheme for n grid points ($n > 7$).

3.4 Solution

(a) The finite difference discretization of the simply supported beam is shown in Fig. 3.17 where at first only the inner nodes will be considered.

It should be noted here that the fourth-order differential equation according to Eq. (3.9) does not allow to account for external single forces (in our case: F_0). The equivalent nodal force R_i is obtained from distributed loads q_X and should not be confused with external single loads. If the derivations should be based on the fourth-order differential equation, then a modeling approach as shown in Fig. 3.18 can be applied. Thus, we understand the single force F_0 as the integral value of a distributed load q_0, which is acting over a length of ΔX. Obviously, this is no more exactly the same load case as shown in Fig. 3.16a. Nevertheless, it allows us to proceed the derivations based on the fourth-order differential equation.

Evaluation of the finite difference approximation of the fourth-order differential equation according to Eq. (3.9) at the inner nodes $i = 2, \ldots, 6$ gives:

Fig. 3.18 Modeling
approach to consider a single
force in the fourth-order
differential equation based
on n grid points

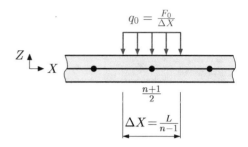

$$\text{node 2: } \frac{EI_Y}{\Delta X^3}\,(u_4 - 4u_3 + 6u_2 - 4u_1 + u_0) = 0\,, \tag{3.97}$$

$$\text{node 3: } \frac{EI_Y}{\Delta X^3}\,(u_5 - 4u_4 + 6u_3 - 4u_2 + u_1) = 0\,, \tag{3.98}$$

$$\text{node 4: } \frac{EI_Y}{\Delta X^3}\,(u_6 - 4u_5 + 6u_4 - 4u_3 + u_2) = -F_0\,(= -q_0\Delta X)\,, \tag{3.99}$$

$$\text{node 5: } \frac{EI_Y}{\Delta X^3}\,(u_7 - 4u_6 + 6u_5 - 4u_4 + u_3) = 0\,, \tag{3.100}$$

$$\text{node 6: } \frac{EI_Y}{\Delta X^3}\,(u_8 - 4u_7 + 6u_6 - 4u_5 + u_4) = 0\,. \tag{3.101}$$

It should be noted here that Eq. (3.99), i.e., the equation with the gray background, is not affected by any boundary or fictitious nodes. This equation will help us later to construct a scheme for a larger number of nodes ($n > 7$). The vertical displacement is zero at both ends and it can be immediately concluded that $u_1 = u_7 = 0$. The fictitious nodes $i = 0$ and $i = 8$ outside the domain can be eliminated based on the boundary condition that the moment must be equal to zero at the supports, i.e. $M_Y(X = 0) = M_Y(X = L) = 0$, see example Problem 3.1. This gives $u_0 = -u_2$ and $u_8 = -U_6$.

Thus, Eqs. (3.97)–(3.101) can be rearranged to give under consideration of the values of the boundary and fictitious nodes

$$\text{node 2: } 5u_2 - 4u_3 + u_4 + 0 + 0 = 0\,, \tag{3.102}$$

$$\text{node 3: } -4u_2 + 6u_3 - 4u_4 + u_5 + 0 = 0\,, \tag{3.103}$$

$$\text{node 4: } u_2 - 4u_3 + 6u_4 - 4u_5 + u_6 = -\frac{\Delta X^3 F_0}{EI_Y}\,, \tag{3.104}$$

$$\text{node 5: } 0 + u_3 - 4u_4 + 6u_5 - 4u_6 = 0,\tag{3.105}$$

$$\text{node 6: } 0 + 0 + u_4 - 4u_5 + 5u_5 = 0,\tag{3.106}$$

or in matrix notation as:

$$
\begin{bmatrix}
5 & -4 & 1 & 0 & 0 \\
-4 & 6 & -4 & 1 & 0 \\
1 & -4 & 6 & -4 & 1 \\
0 & 1 & -4 & 6 & -4 \\
0 & 0 & 1 & -4 & 6
\end{bmatrix}
\begin{bmatrix}
u_2 \\ u_3 \\ u_4 \\ u_5 \\ u_6
\end{bmatrix}
= -\frac{\Delta X^3 F_0}{EI_Y}
\begin{bmatrix}
0 \\ 0 \\ 1 \\ 0 \\ 0
\end{bmatrix}.
\tag{3.107}
$$

The solution of this linear system of equations gives the unknown nodal values as:

$$
\begin{bmatrix}
u_2 \\ u_3 \\ u_4 \\ u_5 \\ u_6
\end{bmatrix}
= -\frac{F_0 L^3}{EI_Y}
\begin{bmatrix}
\frac{1}{96} \\ \frac{1}{54} \\ \frac{19}{864} \\ \frac{1}{54} \\ \frac{1}{96}
\end{bmatrix},
\tag{3.108}
$$

and the relative error in the middle of the beam is obtained as:

$$\text{relative error} = \frac{\frac{19}{864} - \frac{1}{48}}{\frac{1}{48}} \times 100 = 5.56\%.\tag{3.109}$$

From the above calculations, it is easy to derive a general scheme for n nodes ($n > 7$). For simplicity, it is advised to keep a node at $X = L/2$, i.e., the location where the external load is applied to the structure. In generalization of Eq. (3.107), the following scheme can be proposed:

$$
\begin{bmatrix}
5 & -4 & 1 & 0 & 0 & 0 & 0 & 0 & 0 & \cdots & 0 \\
-4 & 6 & -4 & 1 & 0 & 0 & 0 & 0 & 0 & \cdots & 0 \\
1 & -4 & 6 & -4 & 1 & 0 & 0 & 0 & 0 & \cdots & 0 \\
0 & 1 & -4 & 6 & -4 & 1 & 0 & 0 & 0 & \cdots & 0 \\
\vdots & \cdots & & & & & & & \cdots & & \vdots \\
0 & \cdots & 0 & 1 & -4 & 6 & -4 & 1 & 0 & \cdots & 0 \\
\vdots & \cdots & & & & & & & \cdots & & \vdots \\
0 & \cdots & 0 & 0 & 0 & 1 & -4 & 6 & -4 & 1 & 0 \\
0 & \cdots & 0 & 0 & 0 & 0 & 1 & -4 & 6 & -4 & 1 \\
0 & \cdots & 0 & 0 & 0 & 0 & 0 & 1 & -4 & 6 & -4 \\
0 & \cdots & 0 & 0 & 0 & 0 & 0 & 0 & 1 & -4 & 5
\end{bmatrix}
\begin{bmatrix}
u_2 \\ u_3 \\ u_4 \\ u_5 \\ \vdots \\ u_{\frac{n+1}{2}} \\ \vdots \\ u_{n-4} \\ u_{n-3} \\ u_{n-2} \\ u_{n-1}
\end{bmatrix}
= -\frac{\Delta X^3 F_0}{EI_Y}
\begin{bmatrix}
0 \\ 0 \\ 0 \\ 0 \\ \vdots \\ 1 \\ \vdots \\ 0 \\ 0 \\ 0 \\ 0
\end{bmatrix},
\tag{3.110}
$$

Table 3.5 Values of the internal bending moment at the grid points (see Fig. 3.17)

Grid point	Coordinate X	$M_Y(X)$
1	0	0
2	$\frac{L}{6}$	$-\frac{F_0 L}{12}$
3	$\frac{L}{3}$	$-\frac{F_0 L}{6}$
4	$\frac{L}{2}$	$-\frac{F_0 L}{4}$
5	$\frac{2L}{3}$	$-\frac{F_0 L}{6}$
6	$\frac{5L}{6}$	$-\frac{F_0 L}{12}$
7	L	0

where $\Delta X = \frac{L}{n-1}$ for equidistant spacing.

Let us look now on the second approach which is based on the second-order partial differential equation for the moment distribution, see Table 3.1, 3rd form of the PDE. Using a centered difference scheme ($O(\Delta X^2)$) for the second-order derivative (see Table 1.1), one can state the following approximation:

$$EI_Y \frac{\mathrm{d}^2 u}{\mathrm{d}X^2} = EI_Y \frac{u_{i+1} - 2u_i + u_{i-1}}{\Delta X^2} = -M_Y(X). \tag{3.111}$$

Before further evaluating the last equation, let us look on the function of the moment distribution. The moment balance between the external load and the internal bending moment gives finally the following relation ($0 \le X \le L/2$):

$$M_Y(X) = -\frac{F_0}{2} X, \tag{3.112}$$

where the values at the seven grid points are summarized in Table 3.5.

Evaluation of the finite difference approximation of the second-order differential equation according to Eq. (3.111) at the inner nodes $i = 2, \ldots, 6$ gives:

$$\text{node 2: } \frac{EI_Y}{\Delta X^2} (u_3 - 2u_2 + u_1) = \frac{F_0 L}{12}, \tag{3.113}$$

$$\text{node 3: } \frac{EI_Y}{\Delta X^3} (u_4 - 2u_3 + u_2) = \frac{F_0 L}{6}, \tag{3.114}$$

$$\text{node 4: } \frac{EI_Y}{\Delta X^2} (u_5 - 2u_4 + u_3) = \frac{F_0 L}{4}, \tag{3.115}$$

$$\text{node 5: } \frac{E I_Y}{\Delta X^2} (u_6 - 2u_5 + u_4) = \frac{F_0 L}{6}, \tag{3.116}$$

$$\text{node 6: } \frac{E I_Y}{\Delta X^2} (u_7 - 2u_6 + u_5) = \frac{F_0 L}{12}. \tag{3.117}$$

It should be noted here that Eqs. (3.114)–(3.116), i.e., the equations with the gray background, are not affected by any boundary or fictitious nodes. These equations will help us later to construct a scheme for a larger number of nodes ($n > 5$). The vertical displacement is zero at both ends and it can be immediately concluded that $u_1 = u_7 = 0$. Thus, the system of equations can be written in matrix notation as follows:

$$\begin{bmatrix} -2 & 1 & 0 & 0 & 0 \\ 1 & -2 & 1 & 0 & 0 \\ 0 & 1 & -2 & 1 & 0 \\ 0 & 0 & 1 & -2 & 1 \\ 0 & 0 & 0 & 1 & -2 \end{bmatrix} \begin{bmatrix} u_2 \\ u_3 \\ u_4 \\ u_5 \\ u_6 \end{bmatrix} = \frac{\Delta X^2 F_0 L}{E I_Y} \begin{bmatrix} \frac{1}{12} \\ \frac{1}{6} \\ \frac{1}{4} \\ \frac{1}{6} \\ \frac{1}{12} \end{bmatrix}. \tag{3.118}$$

The solution of this linear system of equations gives the unknown nodal values as:

$$\begin{bmatrix} u_2 \\ u_3 \\ u_4 \\ u_5 \\ u_6 \end{bmatrix} = -\frac{F_0 L^3}{E I_Y} \begin{bmatrix} \frac{1}{96} \\ \frac{1}{54} \\ \frac{19}{864} \\ \frac{1}{54} \\ \frac{1}{96} \end{bmatrix}, \tag{3.119}$$

and the relative error in the middle of the beam is obtained as:

$$\text{relative error} = \frac{\frac{19}{864} - \frac{1}{48}}{\frac{1}{48}} \times 100 = 5.56\%. \tag{3.120}$$

From the above calculations, it is easy to derive a general scheme for n nodes ($n > 5$). For simplicity, it is advised to keep a node at $X = L/2$, i.e., the location where the external load is applied to the structure. In generalization of Eq. (3.118), the following scheme can be proposed:

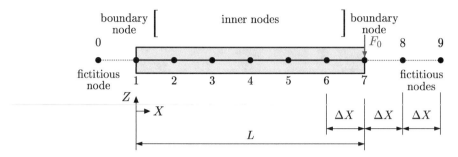

Fig. 3.19 Finite difference discretization of the cantilevered Euler–Bernoulli beam based on seven grid nodes

$$
\begin{bmatrix}
-2 & 1 & 0 & 0 & 0 & 0 & 0 & \cdots & 0 \\
1 & -2 & 1 & 0 & 0 & 0 & 0 & \cdots & 0 \\
0 & 1 & -2 & 1 & 0 & 0 & 0 & \cdots & 0 \\
\vdots & \cdots & & & & & \cdots & & \vdots \\
0 & \cdots & 0 & 1 & -2 & 1 & 0 & \cdots & 0 \\
\vdots & \cdots & & & & & \cdots & & \vdots \\
0 & \cdots & 0 & 0 & 0 & 1 & -2 & 1 & 0 \\
0 & \cdots & 0 & 0 & 0 & 0 & 1 & -2 & 1 \\
0 & \cdots & 0 & 0 & 0 & 0 & 0 & 1 & -2
\end{bmatrix}
\begin{bmatrix}
u_2 \\ u_3 \\ u_4 \\ \vdots \\ u_{\frac{n+1}{2}} \\ \vdots \\ u_{n-3} \\ u_{n-2} \\ u_{n-1}
\end{bmatrix}
= -\frac{\Delta X^2 F_0}{2EI_Y}
\begin{bmatrix}
1\Delta X \\ 2\Delta X \\ 3\Delta X \\ \vdots \\ \frac{n-1}{2}\Delta X \\ \vdots \\ 3\Delta X \\ 2\Delta X \\ 1\Delta X
\end{bmatrix}, \quad (3.121)
$$

where $\Delta X = \frac{L}{n-1}$ for equidistant spacing.

(b) The finite difference discretization of the cantilevered beam is shown in Fig. 3.19 where at first only the inner nodes will be considered.

It should be noted here that the fourth-order differential equation according to Eq. (3.9) does not allow to account for external single forces (in our case: F_0 at the right-hand boundary). The equivalent nodal force R_i is obtained from distributed loads q_X and should not be confused with external single loads. If the derivations should be based on the fourth-order differential equation, then a modeling approach as shown in Fig. 3.20 can be applied. Thus, we understand the single force F_0 as the integral value of a distributed load q_0, which is acting over a length of $\Delta X/2$. Obviously, this is no more exactly the same load case as shown in Fig. 3.16b, i.e., a point load. Nevertheless, it allows us to proceed the derivations based on the fourth-order differential equation.

Evaluation of the finite difference approximation of the nodes with unknown displacements gives similar to part a) the following six equations:

Fig. 3.20 Modeling approach to consider a single force in the fourth-order differential equation based on n grid points

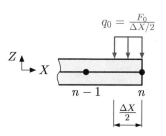

$$\text{node 2: } \frac{EI_Y}{\Delta X^3} (u_4 - 4u_3 + 6u_2 - 4u_1 + u_0) = 0, \tag{3.122}$$

$$\text{node 3: } \frac{EI_Y}{\Delta X^3} (u_5 - 4u_4 + 6u_3 - 4u_2 + u_1) = 0, \tag{3.123}$$

$$\text{node 4: } \frac{EI_Y}{\Delta X^3} (u_6 - 4u_5 + 6u_4 - 4u_3 + u_2) = 0, \tag{3.124}$$

$$\text{node 5: } \frac{EI_Y}{\Delta X^3} (u_7 - 4u_6 + 6u_5 - 4u_4 + u_3) = 0, \tag{3.125}$$

$$\text{node 6: } \frac{EI_Y}{\Delta X^3} (u_8 - 4u_7 + 6u_6 - 4u_5 + u_4) = 0, \tag{3.126}$$

$$\text{node 7: } \frac{EI_Y}{\Delta X^3} (u_9 - 4u_8 + 6u_7 - 4u_6 + u_5) = -F_0. \tag{3.127}$$

It should be noted here that Eqs. (3.124)–(3.125), i.e., the equations with the gray background, are not affected by any boundary or fictitious nodes. These equations will help us later to construct a scheme for a larger number of nodes ($n > 6$). The vertical displacement is zero at left-hand ends and it can be immediately concluded that $u_1 = 0$. Using the condition for the rotation at node 1 ($\frac{du}{dX}\big|_1 = 0$) and the conditions for the internal bending moment ($M_Y|_7 = 0$) and the shear force ($Q_Z|_7 = -F_0$) at node 7 (see example Problem 3.1 for details of the derivations), one can deduct the following additional condition based on centered difference schemes for the derivatives: $u_0 = u_2$, $u_6 = 2u_5 - u_4$, and $u_7 = 4u_5 - 4u_4 + u_3 + 0$.

Thus, Eqs. (3.122)–(3.127) can be rearranged to give under consideration of the values of the boundary and fictitious nodes

$$\text{node 2: } 7u_2 - 4u_3 + u_4 + 0 + 0 + 0 = 0, \tag{3.128}$$

$$\text{node 3: } -4u_2 + 6u_3 - 4u_4 + u_5 + 0 + 0 = 0, \tag{3.129}$$

$$\text{node 4: } u_2 - 4u_3 + 6u_4 - 4u_5 + u_6 + 0 = 0, \tag{3.130}$$

$$\text{node 5: } 0 + u_3 - 4u_4 + 6u_5 - 4u_6 + u_7 = 0, \tag{3.131}$$

$$\text{node } 6: 0 + 0 + u_4 - 4u_5 + 5u_6 - 2u_7 = 0, \tag{3.132}$$

$$\text{node } 7: 0 + 0 + 0 + 2u_5 - 4u_6 + 2u_7 = -\frac{\Delta X^3 F_0}{E I_Y}, \tag{3.133}$$

or in matrix notation:

$$
\begin{bmatrix}
7 & -4 & 1 & 0 & 0 & 0 \\
-4 & 6 & -4 & 1 & 0 & 0 \\
1 & -4 & 6 & -4 & 1 & 0 \\
0 & 1 & -4 & 6 & -4 & 1 \\
0 & 0 & 1 & -4 & 5 & -2 \\
0 & 0 & 0 & 2 & -4 & 2
\end{bmatrix}
\begin{bmatrix}
u_2 \\ u_3 \\ u_4 \\ u_5 \\ u_6 \\ u_7
\end{bmatrix}
= -\frac{\Delta X^{3F_0}}{E I_Y}
\begin{bmatrix}
0 \\ 0 \\ 0 \\ 0 \\ 0 \\ 1
\end{bmatrix}. \tag{3.134}
$$

The solution of this linear system of equations gives the unknown nodal values as:

$$
\begin{bmatrix}
u_2 \\ u_3 \\ u_4 \\ u_5 \\ u_6 \\ u_7
\end{bmatrix}
= -\frac{F_0 L^3}{E I_Y}
\begin{bmatrix}
\frac{1}{144} \\ \frac{11}{432} \\ \frac{23}{432} \\ \frac{19}{216} \\ \frac{55}{432} \\ \frac{73}{432}
\end{bmatrix}, \tag{3.135}
$$

and the relative error right-hand side of the beam is obtained as:

$$\text{relative error} = \frac{\frac{73}{432} - \frac{1}{3}}{\frac{1}{3}} \times 100 = 49.31\%. \tag{3.136}$$

From the above calculations, it is easy to derive a general scheme for n nodes ($n > 7$). In generalization of Eq. (3.134), the following scheme can be proposed:

$$
\begin{bmatrix}
7 & -4 & 1 & 0 & 0 & 0 & 0 & \cdots & 0 \\
-4 & 6 & -4 & 1 & 0 & 0 & 0 & \cdots & 0 \\
1 & -4 & 6 & -4 & 1 & 0 & 0 & \cdots & 0 \\
0 & 1 & -4 & 6 & -4 & 1 & 0 & \cdots & 0 \\
\vdots & \cdots & & & & & & \cdots & \vdots \\
0 & \cdots & 0 & 1 & -4 & 6 & -4 & 1 & 0 \\
0 & \cdots & 0 & 0 & 1 & -4 & 6 & -4 & 1 \\
0 & \cdots & 0 & 0 & 0 & 1 & -4 & 5 & -2 \\
0 & \cdots & 0 & 0 & 0 & 0 & 2 & -4 & 2
\end{bmatrix}
\begin{bmatrix}
u_2 \\ u_3 \\ u_4 \\ u_5 \\ \vdots \\ u_{n-4} \\ u_{n-3} \\ u_{n-2} \\ u_{n-1}
\end{bmatrix}
= -\frac{\Delta X^3 F_0}{E I_Y}
\begin{bmatrix}
0 \\ 0 \\ 0 \\ 0 \\ \vdots \\ 0 \\ 0 \\ 0 \\ 1
\end{bmatrix}, \tag{3.137}
$$

where $\Delta X = \frac{L}{n-1}$ for equidistant spacing.

A different way of solution can be chosen by avoiding the second ($i = 9$) fictitious node at the right-hand end. To this end, a backward finite difference approximation (cf. Table 1.1) can be introduced into the condition for the internal shear force[6] ($Q_Z|_7 = -F_0$) at the right-hand end. Thus, the new equation for node seven can be written as

$$\text{node 7: } 3u_3 - 14u_4 + 24u_5 - 18u_6 + 5u_7 = \frac{2\Delta X^3 F_0}{EI_Y}. \tag{3.138}$$

The corresponding system of equations reads in matrix notation as:

$$
\begin{bmatrix}
7 & -4 & 1 & 0 & 0 & 0 \\
-4 & 6 & -4 & 1 & 0 & 0 \\
1 & -4 & 6 & -4 & 1 & 0 \\
0 & 1 & -4 & 6 & -4 & 1 \\
0 & 0 & 1 & -4 & 5 & -2 \\
0 & 3 & -14 & 24 & -18 & 5
\end{bmatrix}
\begin{bmatrix}
u_2 \\ u_3 \\ u_4 \\ u_5 \\ u_6 \\ u_7
\end{bmatrix}
= \frac{2\Delta X^3 F_0}{EI_Y}
\begin{bmatrix}
0 \\ 0 \\ 0 \\ 0 \\ 0 \\ 1
\end{bmatrix}. \tag{3.139}
$$

The solution of this linear system of equations gives the unknown nodal values as:

$$
\begin{bmatrix}
u_2 \\ u_3 \\ u_4 \\ u_5 \\ u_6 \\ u_7
\end{bmatrix}
= -\frac{F_0 L^3}{EI_Y}
\begin{bmatrix}
\frac{1}{72} \\
\frac{11}{216} \\
\frac{23}{216} \\
\frac{19}{108} \\
\frac{55}{216} \\
\frac{73}{216}
\end{bmatrix}, \tag{3.140}
$$

and the relative error right-hand side of the beam is obtained as:

$$\text{relative error} = \frac{\frac{73}{216} - \frac{1}{3}}{\frac{1}{3}} \times 100 = 1.39\%. \tag{3.141}$$

From the above calculations, it is easy to derive a general scheme for n nodes ($n > 7$). In generalization of Eq. (3.139), the following scheme can be proposed:

---

[6]The value of the internal shear force is now determined based on the real load condition, i.e., Fig. 3.16b, and not based on the modeling approach provided in Fig. 3.20. Thus, we obtain: $Q_Z(X = L) = -F_0$.

Table 3.6 Values of the internal bending moment at the grid points (see Fig. 3.19)

Grid point	Coordinate X	$M_Y(X)$
1	0	$F_0 L$
2	$\frac{L}{6}$	$\frac{5F_0 L}{6}$
3	$\frac{L}{3}$	$\frac{4F_0 L}{6}$
4	$\frac{L}{2}$	$\frac{3F_0 L}{6}$
5	$\frac{2L}{3}$	$\frac{2F_0 L}{6}$
6	$\frac{5L}{6}$	$\frac{1F_0 L}{6}$
7	L	0

$$
\begin{bmatrix}
7 & -4 & 1 & 0 & 0 & 0 & 0 & \cdots & 0 \\
-4 & 6 & -4 & 1 & 0 & 0 & 0 & \cdots & 0 \\
1 & -4 & 6 & -4 & 1 & 0 & 0 & \cdots & 0 \\
0 & 1 & -4 & 6 & -4 & 1 & 0 & \cdots & 0 \\
\vdots & \cdots & & & & & \cdots & & \vdots \\
0 & \cdots & 0 & 1 & -4 & 6 & -4 & 1 & 0 \\
0 & \cdots & 0 & 0 & 1 & -4 & 6 & -4 & 1 \\
0 & \cdots & 0 & 0 & 0 & 1 & -4 & 5 & -2 \\
0 & \cdots & 0 & 0 & 3 & -14 & 24 & -18 & 5
\end{bmatrix}
\begin{bmatrix}
u_2 \\ u_3 \\ u_4 \\ u_5 \\ \vdots \\ u_{n-4} \\ u_{n-3} \\ u_{n-2} \\ u_{n-1}
\end{bmatrix}
= \frac{\Delta X^3 F_0}{E I_Y}
\begin{bmatrix}
0 \\ 0 \\ 0 \\ 0 \\ \vdots \\ 0 \\ 0 \\ 0 \\ 2
\end{bmatrix}, \qquad (3.142)
$$

where $\Delta X = \frac{L}{n-1}$ for equidistant spacing.

Let us look now on the second approach which is based on the second-order partial differential equation for the moment distribution, see Table 3.1, 3rd form of the PDE. Using a centered difference scheme ($O(\Delta X^2)$) for the second-order derivative (see Table 1.1), one can state the following approximation:

$$
E I_Y \frac{d^2 u}{dX^2} = E I_Y \frac{u_{i+1} - 2u_i + u_{i-1}}{\Delta X^2} = -M_Y(X). \qquad (3.143)
$$

Before further evaluating the last equation, let us look on the function of the moment distribution. The moment balance between the external load and the internal bending moment gives finally the following relation ($0 \leq X \leq L$):

$$
M_Y(X) = F_0(L - X), \qquad (3.144)
$$

where the values at the seven grid points are summarized in Table 3.6.

Evaluation of the finite difference approximation of the second-order differential equation according to Eq. (3.143) at the inner nodes $i = 2, \ldots, 6$ gives:

Evaluation of the finite difference approximation of the second-order differential equation according to Eq. (3.143) at the nodes $i = 1, \ldots, 6$ gives[7]:

$$\text{node 1: } \frac{EI_Y}{\Delta X^2} (u_2 - 2u_1 + u_0) = -F_0 L, \tag{3.145}$$

$$\text{node 2: } \frac{EI_Y}{\Delta X^2} (u_3 - 2u_2 + u_1) = -\frac{5F_0 L}{6}, \tag{3.146}$$

$$\text{node 3: } \frac{EI_Y}{\Delta X^2} (u_4 - 2u_3 + u_2) = -\frac{4F_0 L}{6}, \tag{3.147}$$

$$\text{node 4: } \frac{EI_Y}{\Delta X^2} (u_5 - 2u_4 + u_3) = -\frac{3F_0 L}{6}, \tag{3.148}$$

$$\text{node 5: } \frac{EI_Y}{\Delta X^2} (u_6 - 2u_5 + u_4) = -\frac{2F_0 L}{6}, \tag{3.149}$$

$$\text{node 6: } \frac{EI_Y}{\Delta X^2} (u_7 - 2u_6 + u_5) = -\frac{1F_0 L}{6}, \tag{3.150}$$

It should be noted here that Eqs. (3.147)–(3.150), i.e., the equations with the gray background, are not affected by any fictitious nodes or nodes with imposed BCs. These equations will help us later to construct a scheme for a larger number of nodes $(n > 4)$. Equations (3.147)–(3.150) can be written in matrix form as

$$\begin{bmatrix} 1 & 0 & 0 & 0 & 0 & 0 \\ -2 & 1 & 0 & 0 & 0 & 0 \\ 1 & -2 & 1 & 0 & 0 & 0 \\ 0 & 1 & -2 & 1 & 0 & 0 \\ 0 & 0 & 1 & -2 & 1 & 0 \\ 0 & 0 & 0 & 1 & -2 & 1 \end{bmatrix} \begin{bmatrix} u_2 \\ u_3 \\ u_4 \\ u_5 \\ u_6 \\ u_7 \end{bmatrix} = -\frac{\Delta X^2 F_0 L}{EI_Y} \begin{bmatrix} \frac{1}{6} \\ \frac{5}{6} \\ \frac{4}{6} \\ \frac{3}{6} \\ \frac{2}{6} \\ \frac{1}{6} \end{bmatrix}. \tag{3.151}$$

The solution of this linear system of equations gives the unknown nodal values as:

[7] At this point of the derivation, six equations are required. One may consider nodes $i = 2, \ldots, 7$ at a first attempt. However, this would introduce the fictitious node 9. This second fictitious node was eliminated in previous approaches based on the moment equation. Since this approach is based on the PDE in the moment form, the moment equation cannot be used a second time. Thus, the second fictitious cannot be eliminated. In conclusion, one should state the six equations for nodes $i = 1, \ldots, 6$.

$$
\begin{bmatrix} u_2 \\ u_3 \\ u_4 \\ u_5 \\ u_6 \\ u_7 \end{bmatrix} = -\frac{F_0 L^3}{E I_Y} \begin{bmatrix} \frac{1}{36} \\ \frac{17}{216} \\ \frac{4}{27} \\ \frac{25}{108} \\ \frac{35}{108} \\ \frac{91}{216} \end{bmatrix}, \tag{3.152}
$$

and the relative error right-hand side of the beam is obtained as:

$$
\text{relative error} = \frac{\frac{91}{216} - \frac{1}{3}}{\frac{1}{3}} \times 100 = 26.39\% . \tag{3.153}
$$

From the above calculations, it is easy to derive a general scheme for n nodes ($n > 4$). In generalization of Eq. (3.151), the following scheme can be proposed:

$$
\begin{bmatrix}
1 & 0 & 0 & 0 & 0 & \cdots & 0 \\
-2 & 1 & 0 & 0 & 0 & \cdots & 0 \\
1 & -2 & 1 & 0 & 0 & \cdots & 0 \\
0 & 1 & -2 & 1 & 0 & \cdots & 0 \\
\vdots & \cdots & & & & \cdots & \vdots \\
0 & \cdots & 0 & 1 & -2 & 1 & 0 \\
0 & \cdots & 0 & 0 & 1 & -2 & 1
\end{bmatrix}
\begin{bmatrix} u_2 \\ u_3 \\ u_4 \\ u_5 \\ \vdots \\ u_{n-1} \\ u_n \end{bmatrix}
= -F_0 L
\begin{bmatrix}
1 \\
1 - \frac{1}{n-1} \\
1 - \frac{2}{n-1} \\
1 - \frac{3}{n-1} \\
\vdots \\
1 - \frac{n-3}{n-1} \\
1 - \frac{n-2}{n-1}
\end{bmatrix}, \tag{3.154}
$$

where $\Delta X = \frac{L}{n-1}$ for equidistant spacing.

3.5 Finite difference approximation of a stepped cantilevered Euler–Bernoulli beam with a single force based on four or five domain nodes

Given is a stepped Euler–Bernoulli beam of length L with a bending stiffness of $E(2I_Y)$ in the range $0 \leq X \leq L/2$ and a value of $E I_Y$ in the range $L/2 \leq X \leq L$ as shown in Fig. 3.21. The beam is loaded by a single force F_0 at its right-hand boundary. Use four or five domain nodes of equidistant spacing, i.e. $\Delta X = \frac{L}{3}$ or $\Delta X = \frac{L}{4}$, for the finite difference approximation. Use only finite difference approximations of second-order accuracy for the nodal evaluations and boundary conditions. Perform the evaluations (a) starting from Eq. (3.27) and as an alternative considering (b) Eq. (3.25). Determine the vertical displacements at the nodes and compare your result with the analytical solution.

3.5 Solution

(a) The finite difference discretization of the cantilevered beam is shown in Fig. 3.22 for four and five domain nodes.

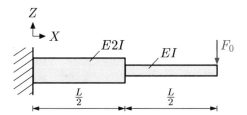

Fig. 3.21 Stepped cantilevered Euler–Bernoulli beam loaded by a single force

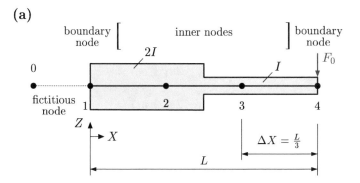

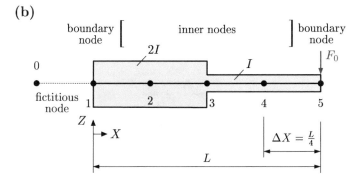

Fig. 3.22 Finite difference discretization of the cantilevered Euler–Bernoulli beam based on **a** four and **b** five grid nodes

Application of Eq. (3.27) requires the distribution of the internal bending moment. The moment balance between the external load and the internal bending moment gives finally the following relation ($0 \leq X \leq L$):

$$M_Y(X) = F_0(L - X),\tag{3.155}$$

where the values at the four grid points are summarized in Table 3.7.

Table 3.7 Values of the internal bending moment at the grid points (see Fig. 3.22a)

Grid point	Coordinate X	$M_Y(X)$
1	0	$F_0 L$
2	$\frac{L}{3}$	$\frac{2F_0 L}{3}$
3	$\frac{2L}{3}$	$\frac{F_0 L}{3}$
4	L	0

Evaluation of the finite difference approximation of the second-order differential equation according to Eq. (3.27) at the nodes $i = 1, \ldots, 3$ gives:

$$\text{node 1:} \quad \frac{E2I_Y}{\Delta X^2} (u_2 - 2u_1 + u_0) = -F_0 L \,, \tag{3.156}$$

$$\text{node 2:} \quad \frac{E2I_Y}{\Delta X^2} (u_3 - 2u_2 + u_1) = -\frac{2F_0 L}{3} \,, \tag{3.157}$$

$$\text{node 3:} \quad \frac{E I_Y}{\Delta X^2} (u_4 - 2u_3 + u_2) = -\frac{F_0 L}{3} \,. \tag{3.158}$$

The vertical displacement is zero at the left-hand boundary because of the fixed support and it follows immediately that $u_1 = 0$ holds. In addition, the rotation is zero at the fixed support, i.e. $\frac{du}{dX}\big|_1 = 0$, and a centered finite difference approach according to Table 1.1 gives the condition $u_0 = u_2$. Introducing these relationships for the support at the left-hand boundary into the system of equations gives:

$$\text{node 1:} \ 2u_2 = -\frac{\Delta X^2 F_0 L}{2E I_Y} \,, \tag{3.159}$$

$$\text{node 2:} \ -2u_2 + u_3 = -\frac{\Delta X^2 F_0 L}{3E I_Y} \,, \tag{3.160}$$

$$\text{node 3:} \ u_2 - 2u_3 + u_4 = -\frac{\Delta X^2 F_0 L}{3E I_Y} \,, \tag{3.161}$$

or in matrix notation:

$$\begin{bmatrix} 2 & 0 & 0 \\ -2 & 1 & 0 \\ 1 & -2 & 1 \end{bmatrix} \begin{bmatrix} u_2 \\ u_3 \\ u_4 \end{bmatrix} = -\frac{\Delta X^2 F_0 L}{E I_Y} \begin{bmatrix} \frac{1}{2} \\ \frac{1}{3} \\ \frac{1}{3} \end{bmatrix} . \tag{3.162}$$

The solution of this linear system of equations gives the unknown nodal values as:

Table 3.8 Values of the internal bending moment at the grid points (see Fig. 3.22b)

Grid point	Coordinate X	$M_Y(X)$
1	0	F_0L
2	$\frac{L}{4}$	$\frac{3F_0L}{4}$
3	$\frac{L}{2}$	$\frac{F_0L}{2}$
4	$\frac{3L}{4}$	$\frac{F_0L}{4}$
5	L	0

$$\begin{bmatrix} u_2 \\ u_3 \\ u_4 \end{bmatrix} = -\frac{F_0L^3}{EI_Y} \begin{bmatrix} \frac{1}{36} \\ \frac{5}{54} \\ \frac{7}{36} \end{bmatrix}, \tag{3.163}$$

and the relative error right-hand side of the beam is obtained as [11]:

$$\text{relative error} = \frac{\frac{7}{36} - \frac{3}{16}}{\frac{3}{16}} \times 100 = 3.70\% . \tag{3.164}$$

Let us now repeat the derivation for five nodes. The evaluation of Eq. (3.155) at the five grid points is summarized in Table 3.8.

Evaluation of the finite difference approximation of the second-order differential equation according to Eq. (3.27) at the nodes $i = 1, \ldots, 4$ gives[8]:

$$\text{node 1: } \frac{E2I_Y}{\Delta X^2}(u_2 - 2u_1 + u_0) = -F_0L , \tag{3.165}$$

$$\text{node 2: } \frac{E2I_Y}{\Delta X^2}(u_3 - 2u_2 + u_1) = -\frac{3F_0L}{4} , \tag{3.166}$$

$$\text{node 3: } \frac{E\left(\frac{2I_Y + I_Y}{2}\right)}{\Delta X^2}(u_4 - 2u_3 + u_2) = -\frac{F_0L}{2} , \tag{3.167}$$

$$\text{node 4: } \frac{EI_Y}{\Delta X^2}(u_5 - 2u_4 + u_3) = -\frac{F_0L}{4} , \tag{3.168}$$

or in matrix notation under consideration of the conditions at the left-hand boundary:

$$\begin{bmatrix} 2 & 0 & 0 & 0 \\ -2 & 1 & 0 & 0 \\ 1 & -2 & 1 & 0 \\ 0 & 1 & -2 & 1 \end{bmatrix} \begin{bmatrix} u_2 \\ u_3 \\ u_4 \\ u_5 \end{bmatrix} = -\frac{\Delta X^2 F_0L}{EI_Y} \begin{bmatrix} \frac{1}{2} \\ \frac{3}{8} \\ \frac{1}{3} \\ \frac{1}{4} \end{bmatrix} . \tag{3.169}$$

---

[8]It should be noted here that the second moment of area is discontinuous at node 3. As a workaround, we simply take the arithmetic mean at this node, i.e., $\frac{2I_Y + I_Y}{2}$.

The solution of this linear system of equations gives the unknown nodal values as:

$$
\begin{bmatrix} u_2 \\ u_3 \\ u_4 \\ u_5 \end{bmatrix} = -\frac{F_0 L^3}{E I_Y} \begin{bmatrix} \frac{1}{64} \\ \frac{7}{128} \\ \frac{11}{96} \\ \frac{73}{384} \end{bmatrix},
\tag{3.170}
$$

and the relative error at the right-hand side of the beam is obtained as [11]:

$$
\text{relative error} = \frac{\frac{73}{384} - \frac{3}{16}}{\frac{3}{16}} \times 100 = 1.39\% .
\tag{3.171}
$$

(b) Application of Eq. (3.25) requires the distribution of the internal shear force $Q_Z(X)$. The vertical force equilibrium between the external load and the internal shear force gives finally the following relation ($0 \le X \le L$):

$$
Q_Z(X) = -F_0 = \text{const.}
\tag{3.172}
$$

Evaluation of the finite difference approximation of the third-order differential equation at the nodes $i = 1, \ldots, 4$ gives[9]:

$$
\text{node 1:} \quad \frac{E2I_Y}{\Delta 2X^3} (-3u_5 + 14u_4 - 24u_3 + 18u_2 - 5u_1) = F_0 ,
\tag{3.173}
$$

$$
\text{node 2:} \quad \frac{E2I_Y}{2\Delta X^3} (u_4 - 2u_3 + 2u_1 - u_0) = F_0 ,
\tag{3.174}
$$

$$
\text{node 3:} \quad E \left(\frac{I_Y - 2I_Y}{2\Delta X} \times \frac{u_4 - 2u_3 + u_2}{\Delta X^2} \right.
$$
$$
\left. + \frac{2I_Y + I_Y}{2(2\Delta X^3)} \times (u_5 - 2u_4 + 2u_2 - u_1) \right) = F_0 ,
\tag{3.175}
$$

$$
\text{node 4:} \quad \frac{E I_Y}{2\Delta X^3} (u_6 - 2u_5 + 2u_3 - u_2) = F_0 ,
\tag{3.176}
$$

Using again the conditions from the left-hand boundary, i.e., $u_1 = 0$ and $u_0 = u_2$ as well as the condition that the moment is zero at the right-hand boundary, i.e. $u_6 = -u_4 + 2u_5$, the matrix notation is obtained as:

$$
\begin{bmatrix} 18 & -24 & 14 & -3 \\ -1 & -2 & 1 & 0 \\ 1 & 1 & -2 & \frac{3}{4} \\ -\frac{1}{2} & 1 & -\frac{1}{2} & 0 \end{bmatrix} \begin{bmatrix} u_2 \\ u_3 \\ u_4 \\ u_5 \end{bmatrix} = \frac{\Delta X^3 F_0}{E I_Y} \begin{bmatrix} 1 \\ 1 \\ 1 \\ 1 \end{bmatrix} .
\tag{3.177}
$$

---

[9]To avoid a second fictitious node at the left-hand boundary, a forward difference approximation is used at node 1. Furthermore, we restrict the derivations to five grid points.

Fig. 3.23 Simply supported
Euler–Bernoulli beam
loaded by two single forces

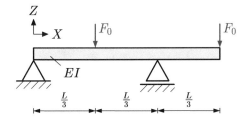

The solution of this linear system of equations gives the unknown nodal values as:

$$
\begin{bmatrix} u_2 \\ u_3 \\ u_4 \\ u_5 \end{bmatrix} = -\frac{F_0 L^3}{E I_Y} \begin{bmatrix} \frac{3}{128} \\ \frac{41}{512} \\ \frac{43}{256} \\ \frac{37}{128} \end{bmatrix},
\tag{3.178}
$$

and the relative error at the right-hand side of the beam is obtained as [11]:

$$
\text{relative error} = \frac{\frac{37}{128} - \frac{3}{16}}{\frac{3}{16}} \times 100 = 54.17\%.
\tag{3.179}
$$

3.6 Example: Finite difference approximation of a simply supported beam with three sections

Given is an simply supported Euler–Bernoulli beam with three different sections as shown in Fig. 3.23. The bending stiffness $E I_Y$ is constant and the entire length is equal to L. The beam is loaded by a single force F_0 at $X = \frac{L}{3}$ and $X = L$. Derive a finite difference approximation based on seven grid points, i.e. an equidistant spacing of $\Delta X = \frac{L}{6}$.

Determine

- the maximum displacement of the beam at the grid points, and
- a schematic sketch of the bending line.
- Compare the results with a modified case where only a single force F_0 is acting at $X = L$.

3.6 Solution

The finite difference discretization based on seven grid points as well as the free-body diagram of the cantilevered beam is shown in Fig. 3.24.

The global moment and vertical force equilibrium yields the reaction forces at the supports as $F_{1Z}^R = 0$ and $F_{5Z}^R = 2F_0$. Thus, we can indicate the internal bending moment functions as follows (see Fig. 3.25a):

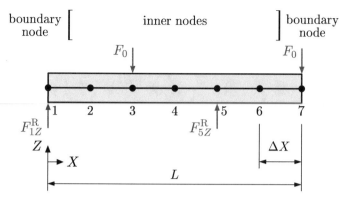

Fig. 3.24 Finite difference discretization and free-body diagram of the cantilevered Euler–Bernoulli beam based on seven grid nodes

$$M_Y(X) = 0 \qquad\qquad \text{for } 0 \le X \le \frac{L}{3}, \tag{3.180}$$

$$M_Y(X) = F_0\left(X - \frac{L}{3}\right) \qquad \text{for } \frac{L}{3} \le X \le \frac{2L}{3}, \tag{3.181}$$

$$M_Y(X) = F_0\,(L - X) \qquad\qquad \text{for } \frac{2L}{3} \le X \le L. \tag{3.182}$$

There are five unknowns to determine, i.e., u_2, u_3, u_4, u_6, u_7, which require to state five equations. Thus, let us state the finite difference approximation of the partial differential equation in the moment formulation for nodes 2, 3, 4, 5, 6. Doing so, no fictitious nodes occur at the boundaries and the evaluation of the inner nodes is sufficient:

$$\text{node 2: } \frac{EI_Y}{\Delta X^2}\,(u_3 - 2u_2 + u_1) = 0, \tag{3.183}$$

$$\text{node 3: } \frac{EI_Y}{\Delta X^2}\,(u_4 - 2u_3 + u_2) = 0, \tag{3.184}$$

$$\text{node 4: } \frac{EI_Y}{\Delta X^2}\,(u_5 - 2u_4 + u_3) = -\frac{F_0 L}{6}, \tag{3.185}$$

$$\text{node 5: } \frac{EI_Y}{\Delta X^2}\,(u_6 - 2u_5 + u_4) = -\frac{F_0 L}{3}, \tag{3.186}$$

$$\text{node 6: } \frac{EI_Y}{\Delta X^2}\,(u_7 - 2u_6 + u_5) = -\frac{F_0 L}{6}, \tag{3.187}$$

or in matrix notation under consideration of both support conditions, i.e., $u_1 = u_5 = 0$:

Fig. 3.25 Schematic representation of the internal bending moment distribution: **a** F_0 at $X = \frac{L}{3}$ and $X = L$ and **b** F_0 at $X = L$

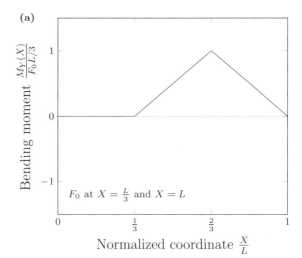

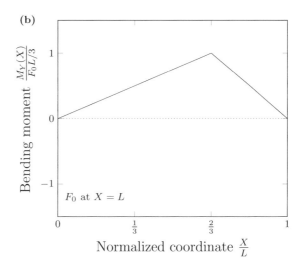

$$
\begin{bmatrix}
-2 & 1 & 0 & 0 & 0 \\
1 & -2 & 1 & 0 & 0 \\
0 & 1 & -2 & 0 & 0 \\
0 & 0 & 1 & 1 & 0 \\
0 & 0 & 0 & -2 & 1
\end{bmatrix}
\begin{bmatrix}
u_2 \\ u_3 \\ u_4 \\ u_6 \\ u_7
\end{bmatrix}
= -\frac{\Delta X^2 F_0 L}{E I_Y}
\begin{bmatrix}
0 \\ 0 \\ \frac{1}{6} \\ \frac{1}{3} \\ \frac{1}{6}
\end{bmatrix} .
\tag{3.188}
$$

The solution of this linear system of equations gives the unknown nodal values as:

$$\begin{bmatrix} u_2 \\ u_3 \\ u_4 \\ u_6 \\ u_7 \end{bmatrix} = -\frac{F_0 L^3}{E I_Y} \begin{bmatrix} \frac{1}{864} \\ \frac{1}{432} \\ \frac{1}{288} \\ -\frac{11}{864} \\ -\frac{13}{432} \end{bmatrix} = -\frac{F_0 L^3}{E I_Y} \begin{bmatrix} 0.0011574074 \\ 0.0023148148 \\ 0.0034722222 \\ -0.012731481 \\ -0.030092593 \end{bmatrix} . \tag{3.189}$$

In the case that only a single external force F_0 is acting at $X = L$, the global moment and vertical force equilibrium yields the reaction forces at the supports as $F_{1Z}^R = -\frac{F_0}{2}$ and $F_{5Z}^R = \frac{3F_0}{2}$. Thus, we can indicate the internal bending moment functions as follows (see Fig. 3.25b):

$$M_Y(X) = \frac{F_0 X}{2} \qquad \text{for } 0 \le X \le \frac{2L}{3}, \tag{3.190}$$

$$M_Y(X) = F_0 (L - X) \qquad \text{for } \frac{2L}{3} \le X \le L. \tag{3.191}$$

Evaluating the finite difference representation for nodes 2, 3, 4, 5, 6, we get:

$$\text{node 2: } \frac{E I_Y}{\Delta X^2} (u_3 - 2u_2 + u_1) = -\frac{F_0 L}{12}, \tag{3.192}$$

$$\text{node 3: } \frac{E I_Y}{\Delta X^2} (u_4 - 2u_3 + u_2) = -\frac{F_0 L}{6}, \tag{3.193}$$

$$\text{node 4: } \frac{E I_Y}{\Delta X^2} (u_5 - 2u_4 + u_3) = -\frac{F_0 L}{4}, \tag{3.194}$$

$$\text{node 5: } \frac{E I_Y}{\Delta X^2} (u_6 - 2u_5 + u_4) = -\frac{F_0 L}{3}, \tag{3.195}$$

$$\text{node 6: } \frac{E I_Y}{\Delta X^2} (u_7 - 2u_6 + u_5) = -\frac{F_0 L}{6}, \tag{3.196}$$

or in matrix notation under consideration of both support conditions, i.e., $u_1 = u_5 = 0$:

$$\begin{bmatrix} -2 & 1 & 0 & 0 & 0 \\ 1 & -2 & 1 & 0 & 0 \\ 0 & 1 & -2 & 0 & 0 \\ 0 & 0 & 1 & 1 & 0 \\ 0 & 0 & 0 & -2 & 1 \end{bmatrix} \begin{bmatrix} u_2 \\ u_3 \\ u_4 \\ u_6 \\ u_7 \end{bmatrix} = -\frac{\Delta X^2 F_0 L}{E I_Y} \begin{bmatrix} \frac{1}{12} \\ \frac{1}{6} \\ \frac{1}{4} \\ \frac{1}{3} \\ \frac{1}{6} \end{bmatrix} . \tag{3.197}$$

The solution of this linear system of equations gives the unknown nodal values as:

Fig. 3.26 Schematic representation of the bending line: **a** F_0 at $X = \frac{L}{3}$ and $X = L$ and **b** F_0 at $X = L$

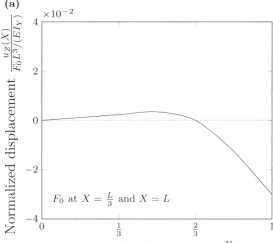

(a)

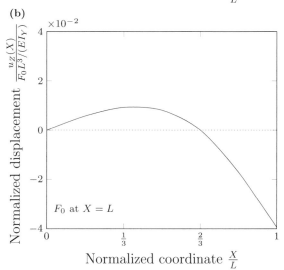

(b)

$$\begin{bmatrix} u_2 \\ u_3 \\ u_4 \\ u_6 \\ u_7 \end{bmatrix} = -\frac{F_0 L^3}{E I_Y} \begin{bmatrix} \frac{5}{864} \\ \frac{1}{108} \\ \frac{7}{864} \\ -\frac{5}{288} \\ -\frac{17}{432} \end{bmatrix} = -\frac{F_0 L^3}{E I_Y} \begin{bmatrix} 0.005787037 \\ 0.0092592593 \\ 0.0081018519 \\ -0.017361111 \\ -0.039351852 \end{bmatrix}. \qquad (3.198)$$

The comparison of the different bending lines is shown in Fig. 3.26. The significant influence of the single force at $X = \frac{l}{3}$ can be clearly seen.

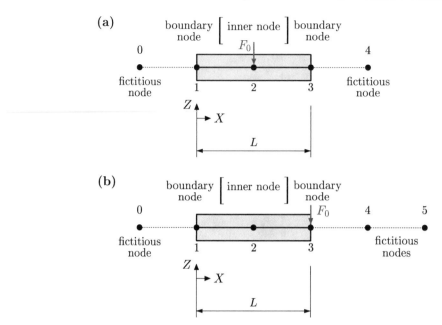

Fig. 3.27 Finite difference approximation based on three domain nodes of **a** a simply supported and **b** a cantilevered Euler–Bernoulli beam loaded by a single force. The schematic sketch of the problem is given in Fig. 3.5

3.5 Supplementary Problems

3.7 Finite difference approximation of a simply supported and cantilevered beam based on three domain nodes

Recalculate example Problem 3.1 based on a discretization of the domain $0 \le X \le L$ with three nodes of equidistant spacing, cf. Fig. 3.27. Compare the results to the approach in example Problem 3.1 which was based on five domain nodes.

3.8 Centered difference approximation of the fourth order derivative

Derive the finite difference approximation for the centered difference scheme of the fourth order derivative where the truncation error is of order ΔX^2 based on Taylor's series expansions around node i for locations $i + 1, i - 1, i + 2$ and $i - 2$. The final result is given in Table 1.1.

3.9 Finite difference approximation of a cantilevered beam based on five domain nodes

Given is a cantilevered Euler–Bernoulli beam of length L with constant bending stiffness EI_Y as shown in Fig. 3.28. The beam is loaded by a single force F_0 at the position $X = \frac{3L}{4}$. Use five domain nodes of equidistant spacing, i.e. $\Delta X = \frac{L}{4}$, for the finite difference approximation. Use only centered difference approximations of second order accuracy for the nodal evaluations and boundary conditions. Determine

Fig. 3.28 Cantilevered
Euler–Bernoulli beam
loaded by a singe force at the
position $X = \frac{3L}{4}$

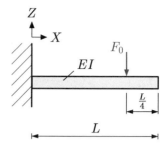

- the vertical displacement at the end of the beam, i.e. $X = L$,
- the analytical solution and
- calculate the relative error between the analytical and finite difference solution.

3.10 Finite difference approximation of a cantilevered beam based on five domain nodes—backward scheme

Reconsider Problem 3.9 and apply a backward finite difference approximation of the third order derivative where the truncation error is of order ΔX^2 to formulate the equilibrium between the internal shear force and external load:

(a) at the free end, $X = L$,
(b) at the loading point, $X = \frac{3L}{4}$.

3.11 Finite difference approximation of a cantilevered beam based on five domain nodes—conversion of tip load into distributed load

Reconsider example Problem 3.1 where a cantilevered beam loaded by a single force was investigated, see Fig. 3.29a. For five domain nodes and a centered difference scheme, a relative error of 54.689% for the vertical deformation at $X = L$ was obtained. In order to improve the centered difference approach, the single force F_0 can be converted in an equivalent distributed load $q_0 = F_0/\Delta X$ between the two last nodes, see Fig. 3.29b. Calculate the relative error of the displacement at $X = L$.

3.12 Finite difference approximation of a simply supported beam based on five domain nodes – bending moment approach

Given is a simply supported Euler–Bernoulli beam of length L with constant bending stiffness EI_Y as shown in Fig. 3.30. The beam is loaded by a constant distributed load q_0. Use five domain nodes of equidistant spacing, i.e. $\Delta X = \frac{L}{4}$, for the finite difference approximation. Use the bending differential equation in the form of the moment, i.e. $EI_Y \frac{\mathrm{d}^2 u_Z}{\mathrm{d}X^2} = -M_Y(X)$, to determine

- the vertical displacement in the middle of the beam, i.e. $X = \frac{L}{2}$,
- the analytical solution and
- calculate the relative error between the analytical and finite difference solution.

(a)

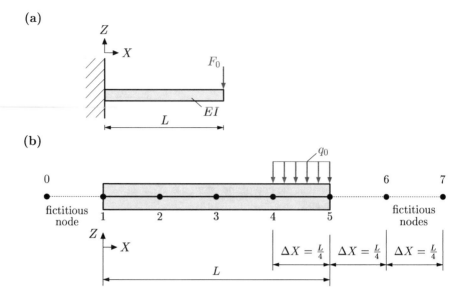

(b)

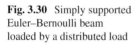

Fig. 3.29 Finite difference approximation of a cantilevered beam based on five domain nodes— conversion of tip load into distributed load

Fig. 3.30 Simply supported Euler–Bernoulli beam loaded by a distributed load

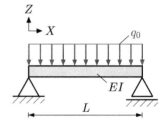

3.13 Finite difference approximation of a simply supported beam under pure bending

Given is a simply supported Euler–Bernoulli beam of length L with constant bending stiffness EI_Y as shown in Fig. 3.31. The beam is loaded by single moments M_0 at its boundaries. Use (a) five and (b) eleven domain nodes of equidistant spacing, i.e. $\Delta X = \frac{L}{4}$ and $\Delta X = \frac{L}{10}$, for the finite difference approximation. Use the bending differential equation in the form of the distributed load and the moment, i.e. $EI_Y \frac{d^4 u_Z}{dX^4} = q_Z(X)$ and $EI_Y \frac{d^2 u_Z}{dX^2} = -M_Y(X)$, to determine

- the vertical displacement at the nodes,
- the analytical solution and
- calculate the relative error between the analytical and finite difference solution for the displacement in the middle of the beam, i.e. $X = \frac{L}{2}$.

Fig. 3.31 Simply supported
Euler–Bernoulli beam under
pure bending

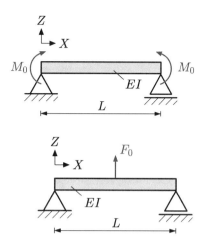

Fig. 3.32 Simply supported
Euler–Bernoulli beam loaded
by a centered single force

3.14 Finite difference approximation of a simply supported beam—displacement, bending moment and shear force distribution

Given is a simply supported Euler–Bernoulli beam of length L with constant bending stiffness $E I_Y$ as shown in Fig. 3.32. The beam is loaded by a single force F_0 at $X = \frac{L}{2}$. Use five domain nodes of equivalent spacing, i.e. $\Delta X = \frac{L}{4}$, for the finite difference approximation. Use the fourth-order bending differential equation (3.7) under consideration of difference schemes of second order accuracy to determine

- the vertical displacement at each node,
- the bending moment at each node,
- the shear force at each node,
- the relative error between the FD and analytical solution at each node in regards to the displacement, moment and shear force.

3.15 Finite difference approximation of a simply supported beam on elastic foundation based on five domain nodes

Given is a simply supported Euler–Bernoulli beam of length L on an elastic foundation as shown in Fig. 3.33. The bending stiffness $E I_Y$ and the elastic foundation modulus k are constant. Use five domain nodes of equidistant spacing, i.e. $\Delta X = \frac{L}{4}$, for the finite difference approximation to determine:

- The vertical displacement in the middle of the beam, i.e. $X = \frac{L}{2}$.
- Compare the finite difference approximation with the analytical solution for the case $k = 4$, $E I_Y = 1$ and $L = 1$.

3.16 Finite difference approximation of a simply supported beam with varying bending stiffness – constant distributed load

Given is a simply supported Euler–Bernoulli beam of length L with varying bending stiffness $E I_Y(X) = \frac{E I_0}{1+(2X/L-1)^2}$ as shown in Fig. 3.34. The beam is loaded by a

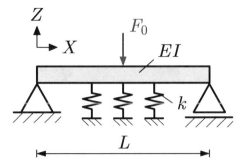

Fig. 3.33 Simply supported Euler–Bernoulli beam on elastic foundation loaded by a single force

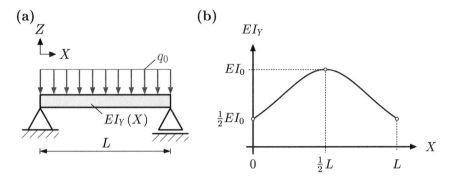

Fig. 3.34 a Simply supported Euler–Bernoulli beam loaded by a distributed load; **b** function of the varying bending stiffness $EI_Y = EI_Y(X)$

constant distributed load q_0. Use five domain nodes of equidistant spacing, i.e. $\Delta X = \frac{L}{4}$, for the finite difference approximation. Use the bending differential equation in the form of the moment, i.e. $EI_Y \frac{d^2 u_Z}{dX^2} = -M_Y(X)$, to determine

- the vertical displacement in the middle of the beam, i.e. $X = \frac{L}{2}$,
- the analytical solution and
- calculate the relative error between the analytical and finite difference solution.

3.17 Finite difference approximation of a simply supported beam with varying bending stiffness – constant bending moment

Given is a simply supported Euler–Bernoulli beam of length L with varying bending stiffness $EI_Y(X) = \frac{EI_0}{1+(2X/L-1)^2}$ as shown in Fig. 3.35. The beam is loaded by single moments M_0 at its ends. Use five domain nodes of equidistant spacing, i.e. $\Delta X = \frac{L}{4}$, for the finite difference approximation. Use the bending differential equation in the form of the moment, i.e. $EI_Y \frac{d^2 u_Z}{dX^2} = -M_Y$, to determine

- the vertical displacement in the middle of the beam, i.e. $X = \frac{L}{2}$,
- the analytical solution and

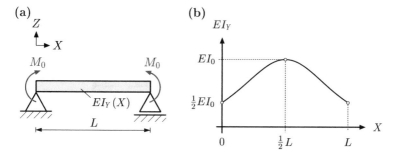

Fig. 3.35 a Simply supported Euler–Bernoulli beam loaded by single moments; **b** function of the varying bending stiffness $EI_Y = EI_Y(X)$

Fig. 3.36 Simply supported Euler–Bernoulli beam loaded by a linearly distributed load

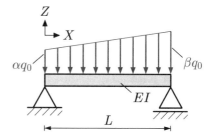

- calculate the relative error between the analytical and finite difference solution.

3.18 Finite difference approximation of a simply supported beam loaded by a linearly distributed load

Given is a simply supported Euler–Bernoulli beam as shown in Fig. 3.36. The bending stiffness EI_Y is constant and the length is equal to L. The simply supported beam is loaded by a linearly varying distributed load $q_Z(X)$ (the boundary values are given as a function of the scalar parameters α and β). Derive a finite difference approximation based on five grid points, i.e. an equidistant spacing of $\Delta X = \frac{L}{4}$. Use centered difference schemes where the truncation error is of order ΔX^2. Determine the displacements at the grid points.

3.19 Finite difference approximation of a stepped cantilevered Euler–Bernoulli beam with two single forces based on five domain nodes

Given is a stepped Euler–Bernoulli beam of length L with a bending stiffness of $E(2I_Y)$ in the range $0 \leq X \leq L/2$ and a value of EI_Y in the range $L/2 \leq X \leq L$ as shown in Fig. 3.37. The beam is loaded by a single force F_0 at $X = \frac{L}{2}$ and a single force F_0 at its right-hand boundary. Use five domain nodes of equidistant spacing, i.e. $\Delta X = \frac{L}{4}$, for the finite difference approximation. Use only finite difference approximations of second-order accuracy for the nodal evaluations and boundary conditions. Perform the evaluations starting from Eq. (3.27). Determine the vertical displacements at the nodes and compare your result with the analytical solution.

Fig. 3.37 Stepped
cantilevered Euler–Bernoulli
beam loaded by two single
forces

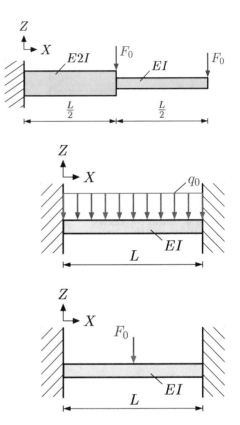

Fig. 3.38 Fixed-ended
Euler–Bernoulli beam
loaded by a distributed load

Fig. 3.39 Fixed-ended
Euler–Bernoulli beam
loaded by a single force

3.20 Finite difference approximation of a fixed-ended beam with a distributed load

Given is a fixed-ended Euler–Bernoulli beam as shown in Fig. 3.38. The bending stiffness $E I_Y$ is constant and the length is equal to L. The fixed-ended beam is loaded by a constant distributed load q_0. Derive finite difference approximations based on five and nine grid points, i.e. an equidistant spacing of $\Delta X = \frac{L}{4}$ or $\Delta X = \frac{L}{8}$. Use centered difference schemes where the truncation error is of order ΔX^2. Determine the displacements at the grid points and compare the results for the different numbers of grid points.

3.21 Finite difference approximation of a fixed-ended beam with a single load

Given is a fixed-ended Euler–Bernoulli beam as shown in Fig. 3.39. The bending stiffness $E I_Y$ is constant and the length is equal to L. The fixed-ended beam is loaded by a single force F_0 in its middle. Derive finite difference approximations based on five and nine grid points, i.e. an equidistant spacing of $\Delta X = \frac{L}{4}$ or $\Delta X = \frac{L}{8}$. Use centered difference schemes where the truncation error is of order ΔX^2. Determine the displacements at the grid points and compare the results for the different numbers of grid points.

Fig. 3.40 Cantilevered
Euler–Bernoulli beam with
prescribed tip displacement

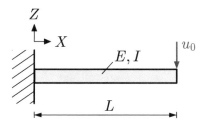

3.22 Finite difference approximation of a cantilevered beam based on five domain nodes with imposed tip displacement

Given is a cantilevered Euler–Bernoulli beam of length L with constant bending stiffness EI_Y as shown in Fig. 3.40. The beam is subjected to an enforced displacement $-u_0$ at the position $X = L$. Use five domain nodes of equidistant spacing, i.e. $\Delta X = \frac{L}{4}$, for the finite difference approximation. Use only centered difference approximations of second order accuracy for the nodal evaluations and boundary conditions. Determine the vertical displacements of the domain nodes.

References

1. Altenbach H, Öchsner A (eds) (2020) Encyclopedia of continuum mechanics. Springer, Berlin
2. Beer FP, Johnston ER Jr, DeWolf JT, Mazurek DF (2009) Mechanics of materials. McGraw-Hill, New York
3. Blaauwendraad J (2010) Plates and FEM: surprises and pitfalls. Springer, Dordrecht
4. Boresi AP, Schmidt RJ (2003) Advanced mechanics of materials. Wiley, New York
5. Budynas RG (1999) Advanced strength and applied stress analysis. McGraw-Hill Book, Singapore
6. Gere JM, Timoshenko SP (1991) Mechanics of materials. PWS-KENT Publishing Company, Boston
7. Gould PL (1988) Analysis of shells and plates. Springer, New York
8. Hibbeler RC (2008) Mechanics of materials. Prentice Hall, Singapore
9. Melosh RJ (1961) A stiffness matrix for the analysis of thin plates in bending. J Aerosp Sci 1:34–64
10. Öchsner A (2014) Elasto-plasticity of frame structure elements: modeling and simulation of rods and beams. Springer, Berlin
11. Öchsner A (2020) Computational statics and dynamics: an introduction based on the finite element method. Springer, Singapore
12. Timoshenko S (1940) Strength of materials - part I elementary theory and problems. D. Van Nostrand Company, New York
13. Timoshenko SP, Goodier JN (1970) Theory of elasticity. McGraw-Hill, New York
14. Ventsel E, Krauthammer T (2001) Thin plates and shells: theory, analysis, and applications. Marcel Dekker, New York
15. Winkler E (1867) Die Lehre von der Elasticität und Festigkeit mit besonderer Rücksicht auf ihre Anwendung in der Technik. H. Dominicus, Prag

Chapter 4
Investigation of Timoshenko Beams in the Elastic Range

4.1 The Basics of a Timoshenko Beam

A thick or Timoshenko beam is defined as a long prismatic body whose axial dimension is much larger than its transverse dimensions [13, 16]. This structural member is only loaded perpendicular to its longitudinal body axis by forces (single forces F_Z or distributed loads q_Z) or moments (single moments M_Y or distributed moments m_Y). Perpendicular means that the line of application of a force or the direction of a moment vector forms a right angle with the X-axis, see Fig. 4.1. As a result of this loading, the deformation occurs only perpendicular to its main axis. The formulation is a shear-flexible theory which means that the shear forces contribute to the bending deformation.

Derivations are restricted many times to the following simplifications:

- only applying to straight beams,
- no elongation along the X-axis,
- no torsion around the X-axis,
- deformations in a single plane (here: X-Z), i.e. symmetrical bending,
- infinitesimally small deformations and strains,
- simple cross sections, and
- the material is linear-elastic, i.e., Youngs's modulus E and shear modulus G.

The three basic equations of continuum mechanics, i.e the kinematics relationship, the constitutive law and the equilibrium equation, as well as their combination to the describing partial differential equations are summarized in Table 4.1. It should be noted here that the deflection u_Z and the rotation ϕ_Y are now independent variables and both represented in the coupled differential equations.

Different formulations of the coupled differential equations are collected in Table 4.2 where different types of loadings, geometry and bedding are differentiated. The last case in Table 4.2 refers again to the elastic or Winkler foundation of

© The Author(s), under exclusive license to Springer Nature Switzerland AG 2021
A. Öchsner, *Structural Mechanics with a Pen*,
https://doi.org/10.1007/978-3-030-65892-2_4

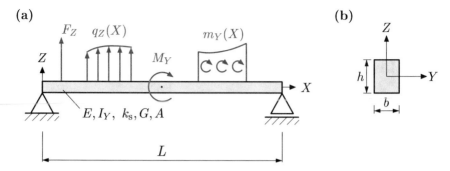

Fig. 4.1 General configuration for Timoshenko beam problems: **a** example of boundary conditions and external loads; **b** cross-sectional area (bending occurs in the X-Z plane)

Table 4.1 Different formulations of the basic equations for a Timoshenko beam (bending in the X-Z plane). e: generalized strains; s: generalized stresses

Specific formulation	General formulation [1]
Kinematics	
$\begin{bmatrix} \frac{du_Z}{dX} + \phi_Y \\ \frac{d\phi_Y}{dX} \end{bmatrix} = \begin{bmatrix} \frac{d}{dX} & 1 \\ 0 & \frac{d}{dX} \end{bmatrix} \begin{bmatrix} u_Z \\ \phi_Y \end{bmatrix}$	$e = \mathcal{L}_1 u$
Constitution	
$\begin{bmatrix} -Q_Z \\ M_Y \end{bmatrix} = \begin{bmatrix} -k_s AG & 0 \\ 0 & EI_Y \end{bmatrix} \begin{bmatrix} \frac{du_Z}{dX} + \phi_Y \\ \frac{d\phi_Y}{dX} \end{bmatrix}$	$s = De$
Equilibrium	
$\begin{bmatrix} \frac{d}{dX} & 0 \\ 1 & \frac{d}{dX} \end{bmatrix} \begin{bmatrix} -Q_Z \\ M_Y \end{bmatrix} + \begin{bmatrix} -q_Z \\ +m_Z \end{bmatrix} = \begin{bmatrix} 0 \\ 0 \end{bmatrix}$	$\mathcal{L}_1^T s + b = 0$
PDE	
$-\frac{d}{dX}\left[k_s GA \left(\frac{du_Z}{dX} + \phi_Y \right) \right] - q_Z = 0$	
$\frac{d}{dX}\left(EI_Y \frac{d\phi_Y}{dX} \right) - k_s GA \left(\frac{du_Z}{dX} + \phi_Y \right) + m_Y = 0,$	$\mathcal{L}_1^T D \mathcal{L}_1 u + b = 0$

a beam, [18]. The elastic foundation modulus k has in the case of beams the unit of force per unit area.

A single-equation description for the Timoshenko beam can be obtained under the assumption of constant material (E, G) and geometrical (I_Y, A, k_s) properties: Rearranging and two-times differentiation of PDEs provided in Table 4.1 gives:

Table 4.2 Different formulations of the partial differential equation for a Timoshenko beam in the X-Z plane (X-axis: right facing; Z-axis: upward facing)

Configuration	Partial differential equation
E, I_Y, A, G, k_s	$EI_Y \dfrac{\mathrm{d}^2\phi_Y}{\mathrm{d}X^2} - k_s GA\left(\dfrac{\mathrm{d}u_Z}{\mathrm{d}X} + \phi_Y\right) = 0$
	$k_s GA\left(\dfrac{\mathrm{d}^2 u_Z}{\mathrm{d}X^2} + \dfrac{\mathrm{d}\phi_Y}{\mathrm{d}X}\right) = 0$
$E(X), I_Y(X)$ $k_s(X), A(X), G(X)$	$\dfrac{\mathrm{d}}{\mathrm{d}X}\left(E(X)I_Y(X)\dfrac{\mathrm{d}\phi_Y}{\mathrm{d}X}\right) -$ $k_s(X)G(X)A(X)\left(\dfrac{\mathrm{d}u_Z}{\mathrm{d}X} + \phi_Y\right) = 0$
	$\dfrac{\mathrm{d}}{\mathrm{d}X}\left[k_s(X)G(X)A(X)\left(\dfrac{\mathrm{d}u_Z}{\mathrm{d}X} + \phi_Y\right)\right] = 0$
$q_Z(X)$	$EI_Y \dfrac{\mathrm{d}^2\phi_Y}{\mathrm{d}X^2} - k_s GA\left(\dfrac{\mathrm{d}u_Z}{\mathrm{d}X} + \phi_Y\right) = 0$
	$k_s GA\left(\dfrac{\mathrm{d}^2 u_Z}{\mathrm{d}X^2} + \dfrac{\mathrm{d}\phi_Y}{\mathrm{d}X}\right) = -q_Z(X)$
$m_Y(X)$	$EI_Y \dfrac{\mathrm{d}^2\phi_Y}{\mathrm{d}X^2} - k_s GA\left(\dfrac{\mathrm{d}u_Z}{\mathrm{d}X} + \phi_Y\right) = -m_Y(X)$
	$k_s GA\left(\dfrac{\mathrm{d}^2 u_Z}{\mathrm{d}X^2} + \dfrac{\mathrm{d}\phi_Y}{\mathrm{d}X}\right) = 0$
$k(X)$	$EI_Y \dfrac{\mathrm{d}^2\phi_Y}{\mathrm{d}X^2} - k_s GA\left(\dfrac{\mathrm{d}u_Z}{\mathrm{d}X} + \phi_Y\right) = 0$
	$k_s GA\left(\dfrac{\mathrm{d}^2 u_Z}{\mathrm{d}X^2} + \dfrac{\mathrm{d}\phi_Y Y}{\mathrm{d}X}\right) = k(X)u_Z$

$$\frac{\mathrm{d}\phi_Y}{\mathrm{d}X} = -\frac{\mathrm{d}^2 u_Z}{\mathrm{d}X^2} - \frac{q_Z}{k_s GA} \,, \tag{4.1}$$

$$\frac{\mathrm{d}^3\phi_Y}{\mathrm{d}X^3} = -\frac{\mathrm{d}^4 u_Z}{\mathrm{d}X^4} - \frac{\mathrm{d}^2 q_Z}{k_s GA \mathrm{d}X^2} \,. \tag{4.2}$$

One-time differentiation of the first PDE provided in Table 4.1 gives:

$$EI_Y \frac{\mathrm{d}^3\phi_Y}{\mathrm{d}X^3} - k_s AG\left(\frac{\mathrm{d}^2 u_Z}{\mathrm{d}X^2} + \frac{\mathrm{d}\phi_Y}{\mathrm{d}X}\right) = 0 \,. \tag{4.3}$$

Inserting Eq. (4.1) into (4.3) and consideration of (4.2) gives finally the following expression:

$$E I_Y \frac{d^4 u_Z(X)}{dX^4} = q_Z(X) - \frac{E I_Y}{k_s A G} \frac{d^2 q_Z(X)}{dX^2} . \tag{4.4}$$

The last equation reduces for shear-rigid beams, i.e. $k_s A G \to \infty$, to the classical Euler–Bernoulli formulation as given in Table 3.2.

Under the assumption of constant material (E, G) and geometric (I_Y, A, k_s) properties, the system of differential equations in Table 4.1 can be solved for constant distributed loads ($q_Z = q_0 = \text{const.}$ and $m_Y = 0$) to obtain the general analytical solution of the problem [15, 16]:

$$u_Z(X) = \frac{1}{E I_Y} \left(\frac{q_0 X^4}{24} + c_1 \frac{X^3}{6} + c_2 \frac{X^2}{2} + c_3 X + c_4 \right) , \tag{4.5}$$

$$\phi_Y(X) = -\frac{1}{E I_Y} \left(\frac{q_0 X^3}{6} + c_1 \frac{X^2}{2} + c_2 X + c_3 \right) - \frac{q_0 X}{k_s A G} - \frac{c_1}{k_s A G} , \tag{4.6}$$

$$M_Y(X) = -\left(\frac{q_0 X^2}{2} + c_1 X + c_2 \right) - \frac{q_0 E I_Y}{k_s A G} , \tag{4.7}$$

$$Q_Z(X) = -(q_0 X + c_1) , \tag{4.8}$$

where the four constants of integration $c_i (i = 1, \ldots, 4)$ must be determined based on the boundary conditions, see Table 4.3.

The internal reactions in a beam become visible if one cuts—at an arbitrary location X—the member in two parts. As a result, two opposite oriented shear forces Q_Z and bending moments M_Y can be indicated. Summing up the internal reactions from both parts must result in zero. Their positive directions are connected with the positive coordinate directions at the positive face (outward surface normal vector parallel to the positive X-axis). This means that at a positive face the positive reactions have the same direction as the positive coordinate axes, see Fig. 4.2.

Once the internal bending moment M_Y is known, the normal stress σ_X can be calculated:

$$\sigma_X(X, Z) = \frac{M_Y(X)}{I_Y} Z(X) = E \frac{d\phi_Y(X)}{dX} Z(X) , \tag{4.9}$$

whereas the shear stress τ_{XZ} is assumed constant over the cross section:

$$\tau_{XZ} = \frac{Q_Z(X)}{A_s} = \frac{Q_Z(X)}{k_s A} = G \gamma_{XZ}(X) . \tag{4.10}$$

Table 4.3 Different boundary conditions and their corresponding reactions for a continuum Timoshenko beam (bending occurs in the X-Z plane)

Case	Boundary condition	Reaction
Z, X (cantilever, fixed at left)	$u_Z(0) = 0,\ \phi_Y(0) = 0$	Z, Y, X; M_Y^R, F^R
(pinned support)	$u_Z(0) = 0,\ M_Y(0) = 0$	F^R
(pinned support)	$u_Z(0) = 0,\ M_Y(0) = 0$	F^R
(fixed/sliding support)	$\phi_Y(0) = 0,\ Q_Z(0) = 0$	M_Y^R
u_0, L	$u_Z(L) = u_0,\ M_Y(L) = 0$	F^R, L
F_0, L	$Q_Z(L) = F_0,\ M_Y(L) = 0$	u, L
ϕ_0, L	$\phi_Y(L) = \phi_0,\ Q_Z(L) = 0$	L, M_Y^R
M_0, L	$M_Y(L) = M_0,\ Q_Z(L) = 0$	L, φ_Y
L	$M_Y(L) = 0,\ Q_Z(L) = 0$	L

Fig. 4.2 Internal reactions for a continuum Timoshenko beam (bending occurs in the X-Z plane)

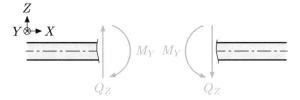

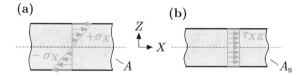

Fig. 4.3 Different stress distributions of a Timoshenko beam with rectangular cross section and linear-elastic material behavior: **a** normal stress and **b** shear stress (bending occurs in the X-Z plane)

In the above equation, the relation between the shear area A_s and the actual cross-sectional area A is referred to as the shear correction factor k_s [4, 6]:

$$k_s = \frac{A_s}{A} .$$ (4.11)

The value of the shear correction factor is, for example, for a circular cross section equal to $\frac{9}{10}$ and for a square cross section equal to $\frac{5}{6}$, see [17].

The relationship between the Young's and shear modulus (see Eqs. (4.9) and (4.10)) is given by [3]:

$$G = \frac{E}{2(1 + \nu)} ,$$ (4.12)

where ν is Poisson's ratio. The graphical representations of the different stress components are shown in Fig. 4.3. The normal stress is, as in the case of the Euler–Bernoulli beam, linearly distributed whereas the shear stress is now assumed to be constant.

If more realistic shear stress distributions are considered, one reaches so-called theories of higher-order [7, 10, 11]. Finally, it should be noted here that the one-dimensional Timoshenko beam theory has its two-dimensional analogy in the form of Reissner–Mindlin plates[1] [2, 5, 8, 12, 14].

4.2   Approximation of the Differential Equations

The consideration of Timoshenko beams requires the simultaneous solution of a coupled system of differential equations as given in Table 4.2. In order to focus on the basic idea of the numerical approach, the derivations in the following are restricted to the case of constant material parameters.

4.1 Example: Finite difference approximation of simply supported beams under different loading conditions

Given is a simply supported Timoshenko beam as shown in Fig. 4.4. The material parameters of the beam are constant and the length is equal to L. The simply supported beam is loaded by a constant distributed load q_0 for case (a) and by a single

---

[1] Also called thick plates.

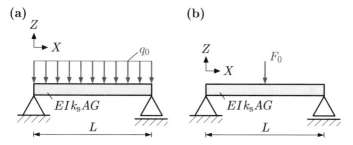

Fig. 4.4 Simply supported Timoshenko beam: **a** distributed load case; **b** single force case

force F_0 for case (b). Use first five domain nodes of equidistant spacing ($\Delta X = \frac{L}{4}$) and then 17 domain nodes of equidistant spacing ($\Delta X = \frac{L}{16}$) for the finite difference approximation.

Determine for both cases

- the general expression for the vertical displacement in the middle of the beam, i.e. $X = \frac{L}{2}$,
- the analytical solution and
- calculate the relative error between the analytical and finite difference solution for a squared cross-sectional area of side length 0.5. Other values are: $E = 10000$; $v = 0.0$; $L = 10$; $q_0 = 1$; $F = 0.1$.

4.1 Solution

The coupled differential equations which describe the problem can be extracted from Table 4.2 as:

$$EI_Z \frac{\mathrm{d}^2\phi_Z}{\mathrm{d}X^2} - k_s GA \left(\frac{\mathrm{d}u_Y}{\mathrm{d}X} + \phi_Z \right) = 0, \tag{4.13}$$

$$k_s GA \left(\frac{\mathrm{d}^2 u_Y}{\mathrm{d}X^2} + \frac{\mathrm{d}\phi_Z}{\mathrm{d}X} \right) = -q_0. \tag{4.14}$$

As can be seen from these two equations, a finite difference approximation is required for first and second order derivatives. These approximations can be taken from Table 1.1 and are given for second order accurate centered difference schemes as[2]:

$$\frac{\mathrm{d}^2\phi_Z}{\mathrm{d}X^2} \approx \frac{\phi_{i+1} - 2\phi_i + \phi_{i-1}}{\Delta X^2} \quad , \quad \frac{\mathrm{d}u_Y}{\mathrm{d}X} \approx \frac{u_{i+1} - u_{i-1}}{2\Delta X} \quad , \tag{4.15}$$

$$\frac{\mathrm{d}^2 u_Y}{\mathrm{d}X^2} \approx \frac{u_{i+1} - 2u_i + u_{i-1}}{\Delta X^2} \quad , \quad \frac{\mathrm{d}\phi_Z}{\mathrm{d}X} \approx \frac{\phi_{i+1} - \phi_{i-1}}{2\Delta X} \quad . \tag{4.16}$$

---

[2]The indices Y and Z will be abandoned in the following to simplify the notation.

Thus, the coupled differential equations given in Eqs. (4.13) and (4.14) can be approximated by the following finite difference scheme:

$$EI_Z \frac{\phi_{i+1} - 2\phi_i + \phi_{i-1}}{\Delta X^2} - k_s GA \left(\frac{u_{i+1} - u_{i-1}}{2\Delta X} + \phi_i \right) = 0, \tag{4.17}$$

$$k_s GA \left(\frac{u_{i+1} - 2u_i + u_{i-1}}{\Delta X^2} + \frac{\phi_{i+1} - \phi_{i-1}}{2\Delta X} \right) = -q_0. \tag{4.18}$$

(a) Distributed load case

Considering five domain nodes (cf. Fig. 3.6), i.e. inner nodes 2, 4 and 5, the finite difference approximations for the three inner nodes can be written as:

$$\text{node 2: } \quad EI_Y \frac{\phi_3 - 2\phi_2 + \phi_1}{\Delta X^2} - k_s GA \left(\frac{u_3 - u_1}{2\Delta X} + \phi_2 \right) = 0, \tag{4.19}$$

$$k_s GA \left(\frac{u_3 - 2u_2 + u_1}{\Delta X^2} + \frac{\phi_3 - \phi_1}{2\Delta X} \right) = -q_0, \tag{4.20}$$

$$\text{node 3: } \quad EI_Y \frac{\phi_4 - 2\phi_3 + \phi_2}{\Delta X^2} - k_s GA \left(\frac{u_4 - u_2}{2\Delta X} + \phi_3 \right) = 0, \tag{4.21}$$

$$k_s GA \left(\frac{u_4 - 2u_3 + u_2}{\Delta X^2} + \frac{\phi_4 - \phi_2}{2\Delta X} \right) = -q_0, \tag{4.22}$$

$$\text{node 4: } \quad EI_Y \frac{\phi_5 - 2\phi_4 + \phi_3}{\Delta X^2} - k_s GA \left(\frac{u_5 - u_3}{2\Delta X} + \phi_4 \right) = 0, \tag{4.23}$$

$$k_s GA \left(\frac{u_5 - 2u_4 + u_3}{\Delta X^2} + \frac{\phi_5 - \phi_3}{2\Delta X} \right) = -q_0. \tag{4.24}$$

Consideration of the boundary conditions for the deflection gives $u(0) = u_1 = 0$ and $u(L) = u_5 = 0$. The unknown rotations at the ends, i.e. ϕ_1 and ϕ_5, can be eliminated from the system of equations by considering the condition for the moments at the ends: $M(0) = 0$ and $M(L) = 0$. From the second constitutive equation provided in Table 4.1, the relationship between bending moment and rotation is given by $M_Y(X) = EI_Y \frac{d\phi_Y}{dX}$ and in order to avoid fictitious nodes, the following forward and backward difference approximations for the first order derivative with second order accuracy can be taken from Table 1.1:

$$\frac{d\phi}{dX} \approx \frac{-3\phi_i + 4\phi_{i+1} - \phi_{i+2}}{2\Delta X} \quad \text{(forward)}, \tag{4.25}$$

$$\frac{d\phi}{dX} \approx \frac{3\phi_i - 4\phi_{i-1} + \phi_{i-2}}{2\Delta X} \quad \text{(backward)}. \tag{4.26}$$

Application of the forward scheme to the left-hand boundary node ($i = 1$) and the backward scheme to the right-hand boundary node ($i = 1$) gives the following two relationships for the elimination of the rotations at the boundaries:

$$\phi_1 = \frac{4}{3}\phi_2 - \frac{1}{3}\phi_3 , \tag{4.27}$$

$$\phi_5 = \frac{4}{3}\phi_4 - \frac{1}{3}\phi_3 . \tag{4.28}$$

Introducing these relationships in the system of equations given by Eqs. (4.19) till (4.24) and considering that the load is opposed to the positive Y-direction, the system of equations can be written in matrix from as given in Eq. (4.30).

The solution of this linear system of equations gives in the middle of the beam under consideration of ($\Delta X = \frac{L}{4}$) the vertical displacement as:

$$u_3 = u\left(\frac{L}{2}\right) = -\frac{\left(32EI_Y + 3\,L^2 k_s AG\right) L^2 q_0}{4k_s AG\left(64EI_Y + L^2 k_s AG\right)} . \tag{4.29}$$

The analytical solution is given in [9] as $u_Z\left(\frac{L}{2}\right) = \frac{5q_0 L^4}{384EI_Y} + \frac{q_0 L^2}{8k_s AG}$ and the relative error results for the given numerical values as -97.208%. Thus, the approximation is not very satisfying and a higher node density is required to improve the results.

$$\begin{bmatrix} 0 & -\frac{2EI_Y}{3\Delta X^2} - k_s AG & -\frac{k_s AG}{2\Delta X} & \frac{2EI_Y}{3\Delta X^2} & 0 & 0 \\[2mm] -\frac{2k_s AG}{\Delta X^2} & -\frac{2k_s AG}{3\Delta X} & \frac{k_s AG}{\Delta X^2} & \frac{2k_s AG}{3\Delta X} & 0 & 0 \\[2mm] \frac{k_s AG}{2\Delta X} & \frac{EI_Y}{\Delta X^2} & 0 & -\frac{2EI_Y}{\Delta X^2} - k_s AG & -\frac{k_s AG}{2\Delta X} & \frac{EI_Y}{\Delta X^2} \\[2mm] \frac{k_s AG}{\Delta X^2} & -\frac{k_s AG}{2\Delta X} & -\frac{2k_s AG}{\Delta X^2} & 0 & \frac{k_s AG}{\Delta X^2} & +\frac{k_s AG}{2\Delta X} \\[2mm] 0 & 0 & \frac{k_s AG}{2\Delta X} & \frac{2EI_Y}{3\Delta X^2} & 0 & -\frac{2EI_Y}{3\Delta X^2} - k_s AG \\[2mm] 0 & 0 & \frac{k_s AG}{\Delta X^2} & -\frac{2k_s AG}{3\Delta X} & -\frac{2k_s AG}{\Delta X^2} & +\frac{2k_s AG}{3\Delta X} \end{bmatrix} \begin{bmatrix} u_2 \\[2mm] \phi_2 \\[2mm] u_3 \\[2mm] \phi_3 \\[2mm] u_4 \\[2mm] \phi_4 \end{bmatrix} = \begin{bmatrix} 0 \\[2mm] q_0 \\[2mm] 0 \\[2mm] q_0 \\[2mm] 0 \\[2mm] q_0 \end{bmatrix} . \tag{4.30}$$

Considering 17 domain nodes, a linear system of dimension 30×30 can be derived based on the 15 inner nodes. The vertical displacement is now as

$$u_9 = u\left(\frac{L}{2}\right) = -\frac{\left(512EI_Y + 53L^2 k_s AG\right)L^2 q_0}{4k_s GA\left(1024EI_Y + L^2 k_s AG\right)} \tag{4.31}$$

obtained and the relative error decreases to -66.348%. As can be seen from this result, a larger number of nodes is required to achieve an acceptable error in the case of Timoshenko beams.

(b) Single force case

In order to account for single forces which are acting on the structure, the system of equations given in Eqs. (4.17) and (4.18) must be multiplied by ΔX to obtain on the right-hand side the expression for the equivalent nodal force. Since this example involves only single *forces*,[3] it would be sufficient to multiply only the second equation by ΔX. Thus, the linear system of equations is obtained for this case as:

$$
\begin{bmatrix}
0 & -\frac{2EI_Y}{3\Delta X^2} - k_s AG & -\frac{k_s AG}{2\Delta X} & \frac{2EI_Y}{3\Delta X^2} & 0 & 0 \\[2ex]
-\frac{2k_s AG}{\Delta X} & -\frac{2k_s AG}{3} & \frac{k_s AG}{\Delta X} & \frac{2k_s AG}{3} & 0 & 0 \\[2ex]
\frac{k_s AG}{2\Delta X} & \frac{EI_Y}{\Delta X^2} & 0 & -\frac{2EI_Y}{\Delta X^2} - k_s AG & -\frac{k_s AG}{2\Delta X} & \frac{EI_Y}{\Delta X^2} \\[2ex]
\frac{k_s AG}{\Delta X} & -\frac{k_s AG}{2} & -\frac{2k_s AG}{\Delta X} & 0 & \frac{k_s AG}{\Delta X} & +\frac{k_s AG}{2} \\[2ex]
0 & 0 & \frac{k_s AG}{2\Delta X} & \frac{2EI_Y}{3\Delta X^2} & 0 & -\frac{2EI_Y}{3\Delta X^2} - k_s AG \\[2ex]
0 & 0 & \frac{k_s AG}{\Delta X} & -\frac{2k_s AG}{3} & -\frac{2k_s AG}{\Delta X} & \frac{2k_s AG}{3}
\end{bmatrix}
\begin{bmatrix}
u_2 \\[2ex] \phi_2 \\[2ex] u_3 \\[2ex] \phi_3 \\[2ex] u_4 \\[2ex] \phi_4
\end{bmatrix}
=
\begin{bmatrix}
0 \\[2ex] 0 \\[2ex] 0 \\[2ex] F_0 \\[2ex] 0 \\[2ex] 0
\end{bmatrix}.
\tag{4.32}
$$

The vertical displacement is for the single load case obtained as

$$u_3 = u\left(\frac{L}{2}\right) = -\frac{\left(32EY_Z + 3L^2 k_s AG\right)LF_0}{2k_s AG\left(64EI_Y + L^2 k_s AG\right)} \tag{4.33}$$

---

[3] In the case of *single* moments, Eq. (4.17) should be multiplied by ΔX.

and the relative error for the given numerical values is equal to -96.514%. Using again 17 domain nodes changes the result in a general form to

$$u_9 = u\left(\frac{L}{2}\right) = -\frac{\left(512EI_Y + 43k_sAGL^2\right)LF_0}{2k_sGA\left(1024EI_Y + L^2k_sAG\right)}, \tag{4.34}$$

which gives a decreased relative error of -65.875%.

4.3   Supplementary Problems

4.2   Finite difference approximation of a beam fixed at both ends

Given is a Timoshenko beam of length L which is fixed at both ends as shown in Fig. 4.5. The beam has constant material parameters and is loaded either by a single force F_0 or a constant distributed load q_0. Use first five domain nodes of equidistant spacing ($\Delta X = \frac{L}{4}$) and then 17 domain nodes of equidistant spacing ($\Delta X = \frac{L}{16}$) for the finite difference approximation to determine

- the general expression for the vertical displacement in the middle of the beam, i.e. $X = \frac{L}{2}$,
- the analytical solution and
- calculate the relative error between the analytical and finite difference solution for a squared cross-sectional area of length 0.5. Other values are: $E = 10000$; $\nu = 0.0$; $L = 10$; $q_0 = 1$; $F_0 = 0.1$.

4.3   Convergence of finite difference approximation of a simply supported Timoshenko beam

Given is a Timoshenko beam of length L which is simply supported as shown in Fig. 4.6. The beam has constant material parameters and is loaded by a constant distributed load q_0. Use the following number of equidistant domain nodes to investigate the convergence of the solution: 5, 13, 23, 33, 53, 73 and 103. The analytical solution can be taken form [9]. Determine

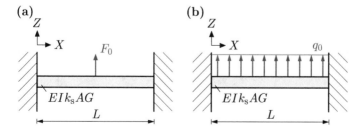

Fig. 4.5 Timoshenko beam fixed at both ends: **a** single force case; **b** distributed load case

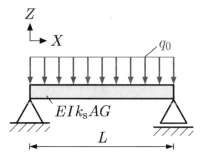

Fig. 4.6 Simply supported Timoshenko beam with distributed load

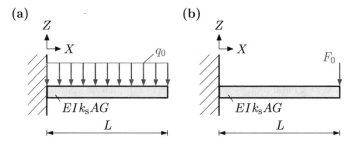

Fig. 4.7 Cantilevered Timoshenko beam with **a** distributed load and **b** single force

- the general expression for the vertical displacement in the middle of the beam, i.e. $X = \frac{L}{2}$,
- a graphical representation of the relative error between the analytical and finite difference solution as a function of the domain node number for a squared cross-sectional area of length 0.5. Other values are: $E = 10000$; $\nu = 0.0$; $L = 10$; $q_0 = 1$.

4.4 Convergence of finite difference approximation of a cantilevered Timoshenko beam

Given is a cantilevered Timoshenko beam of length L as shown in Fig. 4.7. The beam has constant material parameters and is either loaded in the negative Z-direction by a constant distributed load q_0 or a single force F_0. Use the following number of equidistant domain nodes to investigate the convergence of the solution: 5, 13, 23, 33, 53, 73 and 103. The analytical solution can be taken form [9]. Determine

- the general expression for the vertical displacement at the right-hand end of the beam, i.e. $X = L$,
- a graphical representation of the relative error between the analytical and finite difference solution as a function of the domain node number for a squared cross-sectional area of length 0.5. Other values are: $E = 10000$; $\nu = 0.0$; $L = 10$; $q_0 = 1$; $F_0 = 0.1$.

(a) In a first approach, a centered difference approximation should be written for the inner nodes. Thus, a system of equations with the dimension dim. $= 2 \times$ (domain nodes $- 2$) is obtained. The values of u and ϕ at the right-hand boundary $X = L$ should be replaced by appropriate boundary conditions for M_Y and Q_Z based on backward difference schemes of $O(\Delta X^2)$.

(b) In a second approach, write a centered difference approximation which includes in addition the node at $X = L$. Thus, a system of equations with the dimension dim. $= 2 \times$ (domain nodes $- 1$) is obtained. Replace the values for u and ϕ at the fictitious node based on appropriate boundary conditions for M_Y and Q_Z based on centered difference schemes of $O(\Delta X^2)$.

References

1. Altenbach H, Öchsner A (eds) (2020) Encyclopedia of continuum mechanics. Springer, Berlin
2. Blaauwendraad J (2010) Plates and FEM: surprises and pitfalls. Springer, Dordrecht
3. Chen WF, Saleeb AF (1982) Constitutive equations for engineering materials. Volume 1: Elasticity and modelling. Wiley, New York
4. Cowper GR (1966) The shear coefficient in Timoshenko's beam theory. J Appl Mech 33:335–340
5. Gould PL (1988) Analysis of shells and plates. Springer, New York
6. Gruttmann F, Wagner W (2001) Shear correction factors in Timoshenko's beam theory for arbitrary shaped cross-sections. Comput Mech 27:199–207
7. Levinson M (1981) A new rectangular beam theory. J Sound Vib 74:81–87
8. Mindlin RD (1951) Influence of rotary inertia and shear on flexural motions isotropic, elastic plates. J Appl Mech-T ASME 18:1031–1036
9. Öchsner A (2014) Elasto-plasticity of frame structure elements: modeling and simulation of rods and beams. Springer, Berlin
10. Reddy JN (1997) Mechanics of laminated composite plates: theory and analysis. CRC Press, Boca Raton
11. Reddy JN (1997) On locking-free shear deformable beam finite elements. Comput Method Appl M 149:113–132
12. Reissner E (1945) The effect of transverse shear deformation on the bending of elastic plates. J Appl Mech-T ASME 12:A68–A77
13. Timoshenko SP (1921) On the correction for shear of the differential equation for transverse vibrations of prismatic bars. Philos Mag 41:744–746
14. Timoshenko S, Woinowsky-Krieger S (1959) Theory of plates and shells. McGraw-Hill Book Company, New York
15. Wang CM (1995) Timoshenko beam-bending solutions in terms of Euler-Bernoulli solutions. J Eng Mech-ASCE 121:763–765
16. Wang CM, Reddy JN, Lee KH (2000) Shear deformable beams and plates: relationships with classical solution. Elsevier, Oxford
17. Weaver W Jr, Gere JM (1980) Matrix analysis of framed structures. Van Nostrand Reinhold Company, New York
18. Winkler E (1867) Die Lehre von der Elasticität und Festigkeit mit besonderer Rücksicht auf ihre Anwendung in der Technik. H. Dominicus, Prag

Chapter 5
Consideration of Euler–Bernoulli Beams with Plastic Material Behavior

5.1 Basics of the Layered Approach

The derivations of the previous sections will be extended in the following to elasto-plastic material behavior. For simplicity reasons, a simply supported beam under constant moment loading as shown in Fig. 5.1a is considered to explain an approach to consider plasticity for beams. The material is assumed to be linear-elastic/ideal-plastic as shown in Fig. 5.1b [4] and this assumption allows to easily compare the numerical results with the analytical solutions provided in [3].

Let us use five domain nodes ($i = 1, \ldots, 5$) of equidistant spacing ($\Delta X = \frac{L}{4}$) for the finite difference approach as shown in Fig. 5.2a.

As in the case of the elasto-plastic finite *element* approach (see [2, 3, 5]), the external load is now applied in incremental steps of magnitude ΔM, see Fig. 5.2b. The indicated time t is not important since we consider only time-independent material behavior and the expression 'time' can simply be replaced by 'step' or 'load step'. The following derivations do not consider the predictor-corrector scheme as in [2]. A more simplified approach is introduced which is subject to the assumption of monotonic loading. Thus, the cases of unloading, load reversal and cyclic loading will be not covered in the framework of the following derivations.

The major challenge in the case of beams is that the stress state is changing over the height of the beam and a plastic layer is moving inwards with increasing load in the elasto-plastic range. One possible approach for this problem is the so-called layered approach[1] [1] where the cross section at the position of node i is subdivided in a certain numbers of layers (see the grey cross section in Fig. 5.3).

Let us have now a closer look at such a cross section as sown in Fig. 5.4.

The number of the layers ranges between $1 \leq k \leq k_{\max} = 5$ and it is for simplicity assumed that each layer k has the same height Δh. Furthermore, it is assumed that k is an odd number which implies that the center (neural axis) of the beam is located in the middle of layer $\frac{k_{\max}+1}{2}$. The coordinates of the top and bottom face of each layer, i.e. Z^k and Z^{k-1}, can be expressed as

---

[1]The layered approach can be applied in a similar manner in the framework of the finite element method, see Ref. [5].

© The Author(s), under exclusive license to Springer Nature Switzerland AG 2021   103
A. Öchsner, *Structural Mechanics with a Pen*,
https://doi.org/10.1007/978-3-030-65892-2_5

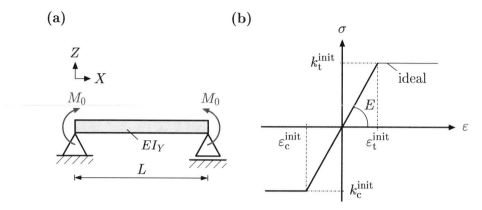

Fig. 5.1 **a** Simply supported Euler–Bernoulli beam under pure bending load; **b** linear-elastic/ideal-plastic material behavior

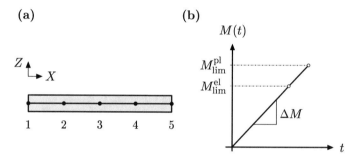

Fig. 5.2 **a** Finite difference discretization of the beam and **b** moment-time function

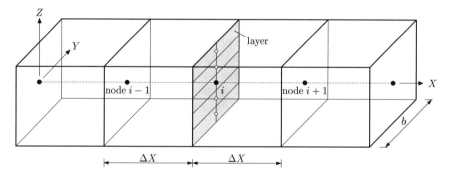

Fig. 5.3 General configuration for the layered beam approach

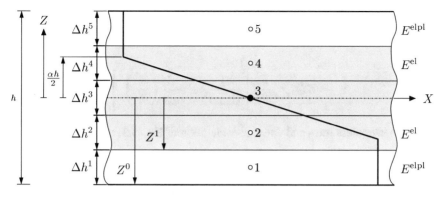

Fig. 5.4 Example of a layered beam with five subdivisions ($k_{max} = 5$)

$$Z^0 = -\frac{h}{2},$$

$$Z^1 = -\frac{h}{2} + \frac{h}{k_{max}},$$

$$Z^2 = -\frac{h}{2} + 2 \times \frac{h}{k_{max}},$$

$$\vdots$$

$$Z^k = -\frac{h}{2} + k \times \frac{h}{k_{max}},$$

$$\vdots$$

$$Z^{k_{max}} = +\frac{h}{2}. \tag{5.1}$$

The height of a layer k ($1 \leq k \leq k_{max}$) can be expressed as

$$\left| \Delta h^k \right| = Z^k - Z^{k-1}, \tag{5.2}$$

and the center coordinate of a layer k is obtained as:

$$Z_c^k = \frac{Z^k + Z^{k-1}}{2} = \frac{1}{2}\left(-h + \frac{(2k-1)h}{k_{max}}\right). \tag{5.3}$$

Thus, the entire bending stiffness of the layered beam is obtained as (cf. Eq. (B.9)):

$$EI_Y = \sum_{k=1}^{k_{max}} E^k \left[\frac{1}{12} b \left(\frac{h}{k_{max}}\right)^3 + b \left(\frac{h}{k_{max}}\right) \left(Z_c^k\right)^2 \right]. \tag{5.4}$$

Looking at the position of a node i, the bending stiffness can be expressed as

$$(EI_Y)_i = \sum_{k=1}^{k_{max}} \frac{E_i^k}{E} \, E \, \underbrace{\frac{1}{12} \, bh^3}_{I_Y} \left[\frac{1}{(k_{max})^3} + \frac{12}{k_{max}} \left(\frac{Z_c^k}{h} \right)^2 \right]_i \tag{5.5}$$

$$= EI_Y \sum_{k=1}^{k_{max}} \frac{E_i^k}{E} \left[\frac{1}{(k_{max})^3} + \frac{12}{k_{max}} \left(\frac{Z_c^k}{h} \right)^2 \right]_i, \tag{5.6}$$

or in dimensionless form and under consideration of Eq. (5.3) as:

$$\frac{(EI_Y)_i}{EI_Y} = \sum_{k=1}^{k_{max}} \frac{E_i^k}{E} \left[\frac{1}{k_{max}^3} + \frac{3}{k_{max}} \left(\frac{2k-1}{k_{max}} - 1 \right)^2 \right]_i = \alpha_i, \tag{5.7}$$

where it should be noted that EI_Y is the bending stiffness of the entire cross section or beam in the pure elastic range and E_i^k is the modulus of the kth layer, i.e. $E_i^k = E$ in the elastic range and $E_i^k = 0$ in the plastic range. Thus, the dimensionless factor is in the pure elastic range $\alpha = 1$ and in the elastic-plastic range $0 \leq \alpha < 1$.

To decide if a layer is now considered as elastic ($E_i^k = E$) or plastic ($E_i^k = 0$), the following assumption is done: If the center of a layer—which is geometrically represented by the coordinate Z_c^k—is in the elastic range, the *entire* layer k is assumed to be elastic. Vice versa, if the center of a layer is in the plastic range, the entire layer k is assumed to be plastic. To develop now an appropriate strategy to decide if the center of a layer is in the pure elastic or elasto-plastic range, one may use the classical assumption that the strain is linearly distributed over the cross section even when the material is in the elasto-plastic range, see Fig. 5.5. Thus, the strain in the center of layer k at node i can be expressed according to the kinematics equation in Table 3.1 as:

$$(\varepsilon_X)_{k,i} = - \left(Z_c^k \times \frac{d^2 u_Z}{dX^2} \right)_i, \tag{5.8}$$

where the second order derivative at node i can be replaced by a centered difference scheme (truncation error of order ΔX^2) as given in Table 1.1. Thus, the discretized form of Eq. (5.8) at node i is given by:

$$(\varepsilon_X)_{k,i} = - \left(Z_c^k \times \frac{u_{i+1} - 2u_i + u_{i-1}}{\Delta X^2} \right)_i. \tag{5.9}$$

It should be noted here that the displacements u_i are calculated for the neutral fiber, i.e. for $Z_c^k = 0$.

The analytical solutions in [3] are presented in a normalized form as

$$\hat{u}_Z(X) = \frac{u_Z(X)}{\frac{M_{lim}^{pl} L^2}{EI_Y}}. \tag{5.10}$$

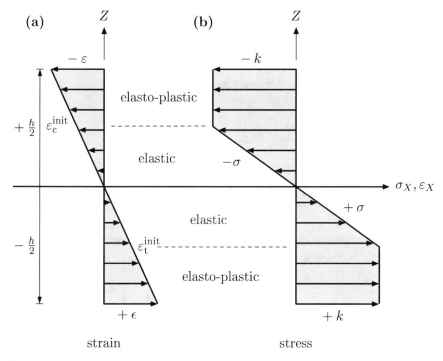

Fig. 5.5 Difference between the **a** strain and **b** stress distribution in the elasto-plastic range for an ideal plastic material. The strain and stress state is uniaxial where only a single normal strain and stress in acting

Introducing the normalized displacements $\hat{u}_Y$ in Eq. (5.9), ones obtains

$$(\varepsilon_X)_{k,i} = -\left(Z_c^k \times \frac{\hat{u}_{i+1} - 2\hat{u}_i + \hat{u}_{i-1}}{\Delta X^2} \times \frac{M_{\text{lim}}^{\text{pl}} L^2}{E I_Y} \right)_i . \qquad (5.11)$$

Based on the relation $M_{\text{lim}}^{\text{pl}} = 3k I_Y / h$ (see [3]) and Hooke's law, the normalizing factor can be expressed as:

$$\frac{M_{\text{lim}}^{\text{pl}} L^2}{E I_Y} = \frac{3k I_Z}{E I_Z h} = \frac{3L^2}{h} \times \frac{k}{E} \overset{\text{Hooke}}{=} \frac{3L^2}{h} \times \varepsilon^{\text{init}} . \qquad (5.12)$$

Finally, one can state the condition for an elastic layer k at node i as:

$$\left| \frac{(\varepsilon_X)_k}{\varepsilon^{\text{init}}} \right|_i = \left| Z_c^k \times \frac{\hat{u}_{i+1} - 2\hat{u}_i + \hat{u}_{i-1}}{\Delta X^2} \times \frac{3L^2}{h} \right|_i \leq 1 \ \text{(elastic)} . \qquad (5.13)$$

If the fraction is larger than one, the layer is assumed to be in the plastic range. In the following, let us look at the entire finite difference solution of the problem. Since we assumed a constant moment loading of the beam (see Fig. 5.1a), one may use the partial differential equation of the problem in the incremental form $EI_Y \frac{d^2 \Delta u_Z}{dX^2} = -\Delta M_Y$ and a centered finite difference approximation to be evaluated at the inner nodes $i = 2, \ldots, 4$ to give[2]:

$$\text{node 2: } \frac{(EI_Y)_2}{\Delta X^2} (\Delta u_3 - 2\Delta u_2 + \Delta u_1) = \Delta M_2 , \tag{5.14}$$

$$\text{node 3: } \frac{(EI_Y)_3}{\Delta X^2} (\Delta u_4 - 2\Delta u_3 + \Delta u_2) = \Delta M_3 , \tag{5.15}$$

$$\text{node 4: } \frac{(EI_Y)_4}{\Delta X^2} (\Delta u_5 - 2\Delta u_4 + \Delta u_3) = \Delta M_4 . \tag{5.16}$$

The last three equations can be represented in matrix form under consideration of the boundary conditions $u_1 = u_5 = 0$ as:

$$\frac{1}{\Delta X^2} \begin{bmatrix} -2(EI_Y)_2 & 1(EI_Y)_2 & 0(EI_Y)_2 \\ 1(EI_Y)_3 & -2(EI_Y)_3 & 1(EI_Y)_3 \\ 0(EI_Y)_4 & 1(EI_Y)_4 & -2(EI_Y)_4 \end{bmatrix} \begin{bmatrix} \Delta u_2 \\ \Delta u_3 \\ \Delta u_4 \end{bmatrix} = \begin{bmatrix} \Delta M_2 \\ \Delta M_3 \\ \Delta M_4 \end{bmatrix} . \tag{5.17}$$

Since the coefficient matrix is now dependent on the solution (i.e. depending on the deformation, the $(EI_Y)_i$ may take different values), a residual form can be written as

$$\begin{bmatrix} r_2 \\ r_3 \\ r_4 \end{bmatrix} = \frac{1}{\Delta X^2} \begin{bmatrix} -2(EI_Y)_2 & 1(EI_Y)_2 & 0(EI_Y)_2 \\ 1(EI_Y)_3 & -2(EI_Y)_3 & 1(EI_Y)_3 \\ 0(EI_Y)_4 & 1(EI_Y)_4 & -2(EI_Y)_4 \end{bmatrix} \begin{bmatrix} \Delta u_2 \\ \Delta u_3 \\ \Delta u_4 \end{bmatrix} - \begin{bmatrix} \Delta M_2 \\ \Delta M_3 \\ \Delta M_4 \end{bmatrix} , \tag{5.18}$$

or in abbreviated form as:

$$r(\Delta u) = K(\Delta u)\Delta u - \Delta F . \tag{5.19}$$

To solve this nonlinear system of equations, a complete Newton–Raphson iteration (iteration index j) can be applied in the following form [2]:

$$\Delta u^{(j+1)} = \Delta u^{(j)} - \left(K_T^{(j)} \right)^{-1} r(\Delta u^{(j)}) , \tag{5.20}$$

where K_T is the so-called tangent stiffness matrix. Equation (5.20) must be iterated as long as, for example, the normalized difference between two consecutive iteration steps is not below a certain threshold. Thus, the iteration can be stopped if the following condition is satisfied:

---

[2]The external moment M_0 given in Fig. 5.1 results in a *negative* internal bending moment distribution: $M_Y < 0$.

$$\sqrt{\frac{\left(\Delta u_2^{(j)} - \Delta u_2^{(j-1)}\right)^2 + \left(\Delta u_3^{(j)} - \Delta u_3^{(j-1)}\right)^2 + \left(\Delta u_4^{(j)} - \Delta u_4^{(j-1)}\right)^2}{\left(\Delta u_2^{(j)}\right)^2 + \left(\Delta u_3^{(j)}\right)^2 + \left(\Delta u_4^{(j)}\right)^2}} \leq t_{\text{end}} ,$$

(5.21)

where t_{end} is a small number. Before applying the iteration scheme of Eq. (5.20), it is appropriate to have a closer look at the tangent stiffness matrix $\boldsymbol{K}_T$. In general, the tangent stiffness matrix can be expressed as [2]

$$\boldsymbol{K}_T = \frac{\partial \boldsymbol{r}(\Delta \boldsymbol{u})}{\partial \Delta \boldsymbol{u}} = \boldsymbol{K} + \frac{\partial \boldsymbol{K}}{\partial \Delta \boldsymbol{u}} \Delta \boldsymbol{u} ,$$

(5.22)

which can be written in our case of three unknowns as:

$$\boldsymbol{K}_T = \underbrace{\boldsymbol{K}}_{3 \times 3} + \underbrace{\left[\frac{\partial \boldsymbol{K}}{\partial \Delta u_2} \Delta \boldsymbol{u} \,\middle|\, \frac{\partial \boldsymbol{K}}{\partial \Delta u_3} \Delta \boldsymbol{u} \,\middle|\, \frac{\partial \boldsymbol{K}}{\partial \Delta u_4} \Delta \boldsymbol{u} \right]}_{3 \times 3 \text{ matrix}},$$

(5.23)

or in components:

$$\boldsymbol{K}_T = \begin{bmatrix} K_{11} & K_{12} & K_{13} \\ K_{21} & K_{22} & K_{23} \\ K_{31} & K_{32} & K_{33} \end{bmatrix} + \left(\begin{bmatrix} \frac{\partial K_{11}}{\partial \Delta u_2} & \frac{\partial K_{12}}{\partial \Delta u_2} & \frac{\partial K_{13}}{\partial \Delta u_2} \\ \frac{\partial K_{21}}{\partial \Delta u_2} & \frac{\partial K_{22}}{\partial \Delta u_2} & \frac{\partial K_{23}}{\partial \Delta u_2} \\ \frac{\partial K_{31}}{\partial \Delta u_2} & \frac{\partial K_{32}}{\partial \Delta u_2} & \frac{\partial K_{33}}{\partial \Delta u_2} \end{bmatrix} \begin{bmatrix} \Delta u_2 \\ \Delta u_3 \\ \Delta u_4 \end{bmatrix} \middle| \right.$$

$$\left. \begin{bmatrix} \frac{\partial K_{11}}{\partial \Delta u_3} & \frac{\partial K_{12}}{\partial \Delta u_3} & \frac{\partial K_{13}}{\partial \Delta u_3} \\ \frac{\partial K_{21}}{\partial \Delta u_3} & \frac{\partial K_{22}}{\partial \Delta u_3} & \frac{\partial K_{23}}{\partial \Delta u_3} \\ \frac{\partial K_{31}}{\partial \Delta u_3} & \frac{\partial K_{32}}{\partial \Delta u_3} & \frac{\partial K_{33}}{\partial \Delta u_3} \end{bmatrix} \begin{bmatrix} \Delta u_2 \\ \Delta u_3 \\ \Delta u_4 \end{bmatrix} \middle| \begin{bmatrix} \frac{\partial K_{11}}{\partial \Delta u_4} & \frac{\partial K_{12}}{\partial \Delta u_4} & \frac{\partial K_{13}}{\partial \Delta u_4} \\ \frac{\partial K_{21}}{\partial \Delta u_4} & \frac{\partial K_{22}}{\partial \Delta u_4} & \frac{\partial K_{23}}{\partial \Delta u_4} \\ \frac{\partial K_{31}}{\partial \Delta u_4} & \frac{\partial K_{32}}{\partial \Delta u_4} & \frac{\partial K_{33}}{\partial \Delta u_4} \end{bmatrix} \begin{bmatrix} \Delta u_2 \\ \Delta u_3 \\ \Delta u_4 \end{bmatrix} \right).$$

(5.24)

The partial derivatives in Eq. (5.24) can be identified under consideration of Eq. (5.17) as partial derivatives of the bending stiffness $(EI_Y)_i$ with respect to one of the unknown displacement increments. Since the second moment of area can be considered in our specific case as a constant, the partial derivatives take the form:

$$\frac{\partial K_{kl}}{\partial \Delta u} \sim \frac{\partial E}{\partial \Delta u} .$$

(5.25)

Looking at Fig. 5.6, one can conclude that these partial derivatives are in the case of an ideal plastic—or even linear hardening—material equal to zero and the tangent stiffness matrix reduces for this special case to the stiffness matrix: $\boldsymbol{K}_T = \boldsymbol{K}$.

Thus, the iteration scheme of Eq. (5.20) reduces in this special case to:

$$\Delta \boldsymbol{u}^{(j+1)} = \Delta \boldsymbol{u}^{(j)} - \left(\boldsymbol{K}^{(j)}\right)^{-1} \left(\boldsymbol{K}(\Delta \boldsymbol{u}) \Delta \boldsymbol{u} - \Delta \boldsymbol{F}\right)^{(j)}$$

(5.26)

$$= \left(\boldsymbol{K}^{(j)}\right)^{-1} \Delta \boldsymbol{F}^{(j)},$$

(5.27)

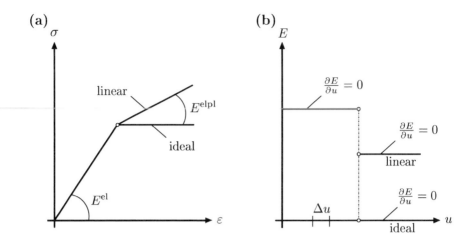

Fig. 5.6 **a** Stress-strain diagram and **b** modulus as a function of displacement

which is known under the expression direct or Picard's iteration [6]. In components, we can write this scheme for our specific case as:

$$
\begin{bmatrix} \Delta u_2 \\ \Delta u_3 \\ \Delta u_4 \end{bmatrix}^{(j+1)} = \frac{\Delta X^2}{4} \begin{bmatrix} \frac{-3}{(EI_Y)_2} & \frac{-2}{(EI_Y)_3} & \frac{-1}{(EI_Y)_4} \\ \frac{-2}{(EI_Y)_2} & \frac{-4}{(EI_Y)_3} & \frac{-2}{(EI_Y)_4} \\ \frac{-1}{(EI_Y)_2} & \frac{-2}{(EI_Y)_3} & \frac{-3}{(EI_Y)_4} \end{bmatrix}^{(j)} \begin{bmatrix} \Delta M_2 \\ \Delta M_3 \\ \Delta M_4 \end{bmatrix}^{(j)}
\tag{5.28}
$$

$$
\stackrel{(5.8)}{=} \frac{M^{\mathrm{pl}}_{\mathrm{lim}} L^2}{64 E I_Y} \begin{bmatrix} \frac{-3}{\alpha_2} & \frac{-2}{\alpha_3} & \frac{-1}{\alpha_4} \\ \frac{-2}{\alpha_2} & \frac{-4}{\alpha_3} & \frac{-2}{\alpha_4} \\ \frac{-1}{\alpha_2} & \frac{-2}{\alpha_3} & \frac{-3}{\alpha_4} \end{bmatrix}^{(j)} \begin{bmatrix} \frac{\Delta M_2}{M^{\mathrm{pl}}_{\mathrm{lim}}} \\ \frac{\Delta M_3}{M^{\mathrm{pl}}_{\mathrm{lim}}} \\ \frac{\Delta M_4}{M^{\mathrm{pl}}_{\mathrm{lim}}} \end{bmatrix}^{(j)} ,
\tag{5.29}
$$

or in normalized form as:

$$
\begin{bmatrix} \Delta \hat{u}_2 \\ \Delta \hat{u}_3 \\ \Delta \hat{u}_4 \end{bmatrix}^{(j+1)} = \frac{1}{\frac{M^{\mathrm{pl}}_{\mathrm{lim}} L^2}{E I_Y}} \begin{bmatrix} \Delta u_2 \\ \Delta u_3 \\ \Delta u_4 \end{bmatrix}^{(j+1)} = \frac{1}{64} \begin{bmatrix} \frac{-3}{\alpha_2} & \frac{-2}{\alpha_3} & \frac{-1}{\alpha_4} \\ \frac{-2}{\alpha_2} & \frac{-4}{\alpha_3} & \frac{-2}{\alpha_4} \\ \frac{-1}{\alpha_2} & \frac{-2}{\alpha_3} & \frac{-3}{\alpha_4} \end{bmatrix}^{(j)} \begin{bmatrix} \frac{\Delta M_2}{M^{\mathrm{pl}}_{\mathrm{lim}}} \\ \frac{\Delta M_3}{M^{\mathrm{pl}}_{\mathrm{lim}}} \\ \frac{\Delta M_4}{M^{\mathrm{pl}}_{\mathrm{lim}}} \end{bmatrix}^{(j)} .
\tag{5.30}
$$

Let us now evaluate this iteration scheme for different ratios of the normalized moment in the pure elastic and the elasto-plastic range and compare these numerical results with the exact analytical solution. As shown in Fig. 5.2a, the beam is sub-

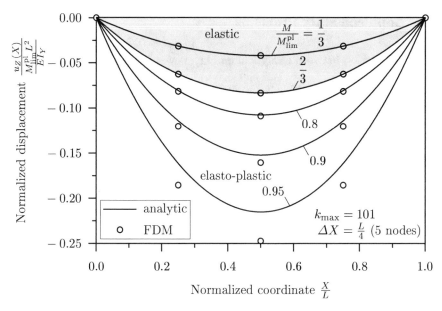

Fig. 5.7 Normalized deformed shape of a simply supported Euler–Bernoulli beam under pure bending load. Comparison between analytical and finite difference solution

divided in five equidistant nodes and the number of layers is taken to be equal to 101 for the following calculations. Since the iteration scheme is quite sensible to the load step, the following increments should be assigned: Between $\frac{M}{M^{\mathrm{pl}}_{\mathrm{lim}}} = \frac{2}{3}$ and 0.8: 10 increments; between $\frac{M}{M^{\mathrm{pl}}_{\mathrm{lim}}} = 0.8$ and 0.9: 10 increments; between $\frac{M}{M^{\mathrm{pl}}_{\mathrm{lim}}} = 0.9$ and 0.95: 10 increments.

The results of the finite difference iterations are compared to the analytical solution in Fig. 5.7. As can be seen from this figure, the elastic finite difference solution is identical to the analytical solution and no incremental load application is required. The plastic range shows, however, an increasing difference between finite difference and analytical solution with increasing plastic deformation. To increase the accuracy in the plastic range, the following three parameters can be modified:

- the total number of nodes, i.e. a finer discretization,
- the number of the layers (k_{max}) over the height of the cross section and
- the size of the load increment ΔM.

The influence of these parameters on the accuracy will be investigated in the scope of supplementary problems.

5.2   Supplementary Problems

5.1 Comparison between analytical and layer-wise integration of the bending stiffness

Given is a segment of an Euler–Bernoulli beam which is divided in k_{max} equidistant layers as shown in Fig. 5.8. The plastic zone extends to the coordinate $Z = \pm\frac{\alpha h}{2}$. Determine the bending stiffness EI_Y based on the exact approach and based on layer-wise integration. Sketch the relative error between both approaches for the specific values $\alpha = \frac{11}{20}$ and $\frac{1}{2}$ in the range $0 \leq k \leq k_{max} = 151$.

5.2 Investigation of the proportions of the relative bending stiffness

The relative bending stiffness within the layered approach can be expressed according to Eq. (5.7) for the pure elastic case as

$$\frac{(EI_Y)_i}{EI_Y} = \sum_{k=1}^{k_{max}} 1 \times \left[\frac{1}{k_{max}^3} + \frac{3}{k_{max}} \left(\frac{2k-1}{k_{max}} - 1 \right)^2 \right]_i \tag{5.31}$$

$$= \sum_{k=1}^{k_{max}} 1 \times [\beta_1 + \beta_2]_i \ , \tag{5.32}$$

where the dimensionless factor β_1 is the proportion of the cross section related to the layer's own coordinate system and β_2 is the contribution from the offset between the coordinate system of the neutral axis and the layer's own coordinate system (see the parallel axis theorem). Some references neglect the contribution of β_1. Thus, determine the absolute difference between β_1, $\beta_2(k = k_{max})$ and $\beta_2\left(k = \frac{k_{max}+3}{2}\right)$ as well as the relative difference between β_1 and both functions of β_2 in the range $1 \leq k_{max} \leq 151$.

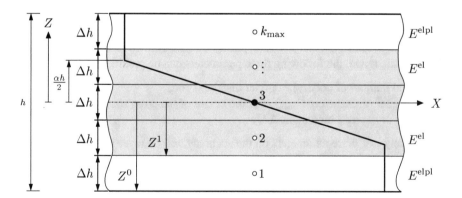

Fig. 5.8 Schematic sketch for the layer-wise integration of the bending stiffness

5.3 Elasto-plastic finite difference solution: influence of layer number

Consider again the problem of the simply supported Euler–Bernoulli beam under constant bending load as shown in Figs. 5.1 and 5.7. The beam should be discretized by five equidistant nodes and the moment should be applied in the range $M/M_{\text{lim}}^{\text{pl}} = 2/3$ and 0.8 and in the range from 0.8 to 0.9 in ten equidistant steps. Calculate the deformation for $M/M_{\text{lim}}^{\text{pl}} = 0.9$ with $k_{\text{max}} = 11, 101, 1001, 10001$ and determine in addition the relative error to the analytical solution.

5.4 Elasto-plastic finite difference solution: influence of load increment

Consider again the problem of the simply supported Euler–Bernoulli beam under constant bending load as shown in Figs. 5.1 and 5.7. The beam should be discretized by five equidistant nodes and the layers are $k_{\text{max}} = 101$. The moment should be applied in the range $M/M_{\text{lim}}^{\text{pl}} = 2/3$ and 0.8 and in the range from 0.8 to 0.9 in 5, 8 or 10 equidistant steps. Calculate the deformation for $M/M_{\text{lim}}^{\text{pl}} = 0.9$ and determine in addition the relative error to the analytical solution in dependence of the size of the load increment.

References

1. Al-Amery RIM, Roberts TM (1990) Nonlinear finite difference analysis of composite beams with partial interaction. Comput Struct 35:81–87
2. Öchsner A, Merkel M (2018) One-dimensional finite elements: an introduction to the FE method. Springer, Cham
3. Öchsner A (2014) Elasto-plasticity of frame structure elements: modeling and simulation of rods and beams. Springer, Berlin
4. Öchsner A (2016) Continuum damage and fracture mechanics. Springer, Singapore
5. Owen DRJ, Hinton E (1980) Finite elements in plasticity: theory and practice. Pineridge Press Limited, Swansea
6. Reddy JN (2004) An introduction to nonlinear finite element analysis. Oxford University Press, Oxford

Chapter 6
Answers to Supplementary Problems

6.1 Answers for Problems from Chap. 1

1.1 Forward difference approximation of the first order derivative
Intermediate result:

$$\left(\frac{du}{dX}\right)_i = \frac{u_{i+1} - u_i}{\Delta X} - \frac{1}{2}\left(\frac{d^2u}{dX^2}\right)_i \Delta X \underbrace{- \frac{1}{6}\left(\frac{d^3u}{dX^3}\right)_i \Delta X^2 - \cdots}_{O(\Delta X^2)} . \qquad (6.1)$$

1.2 Centered difference approximation of the third order derivative
Starting point:

$$\left(\frac{du}{dX}\right)_i = \frac{u_{i+1} - u_{i-1}}{2\Delta X} . \qquad (6.2)$$

Introducing in the above equation the expressions for the second order derivatives at nodes $i + 1$ and $i - 1$, i.e.

$$\left(\frac{d^2u}{dX^2}\right)_{i+1} \approx \frac{u_{i+2} - 2u_{i+1} + u_i}{\Delta X^2} \rightarrow u_{i+1} , \qquad (6.3)$$

$$\left(\frac{d^2u}{dX^2}\right)_{i-1} \approx \frac{u_i - 2u_{i-1} + u_{i-2}}{\Delta X^2} \rightarrow u_{i-1} , \qquad (6.4)$$

gives the derivative of the second order derivative, i.e. the third order derivative, as stated in Table 1.1.

© The Author(s), under exclusive license to Springer Nature Switzerland AG 2021
A. Öchsner, *Structural Mechanics with a Pen*,
https://doi.org/10.1007/978-3-030-65892-2_6

6.2 Answers for Problems from Chap. 2

2.3 Finite difference approximation of a cantilevered rod with distributed load based on five domain nodes

The linear system of equations reads as:

$$
\frac{EA}{\Delta X}
\begin{bmatrix}
2 & -1 & 0 & 0 \\
-1 & 2 & -1 & 0 \\
0 & -1 & 2 & -1 \\
0 & 0 & -1 & 1
\end{bmatrix}
\begin{bmatrix}
u_2 \\ u_3 \\ u_4 \\ u_5
\end{bmatrix}
=
\begin{bmatrix}
\Delta X p_0 \\
\Delta X p_0 \\
\Delta X p_0 \\
\Delta X p_0 \\
\frac{\Delta X p_0}{4}
\end{bmatrix}
, \tag{6.5}
$$

or solved for the nodal unknowns:

$$
\begin{bmatrix}
u_2 \\ u_3 \\ u_4 \\ u_5
\end{bmatrix}
=
\begin{bmatrix}
\frac{13}{4}\frac{\Delta X^2 p_0}{EA} \\
\frac{11}{2}\frac{\Delta X^2 p_0}{EA} \\
\frac{27}{4}\frac{\Delta X^2 p_0}{EA} \\
\frac{7}{1}\frac{\Delta X^2 p_0}{EA}
\end{bmatrix}
. \tag{6.6}
$$

The analytical solution can be taken from [1] as $u_5 = \frac{p_0 L^2}{2EA}$ whereas the finite difference solution gives $u_5 = \frac{7}{16}\frac{p_0 L^2}{EA} \approx 0.4375 \times \frac{p_0 L^2}{EA}$.

2.4 Refined finite difference approximation of a cantilevered rod with distributed load

In generalization of Eq. (6.5), the following scheme can be proposed:

$$
\frac{EA}{\Delta X}
\begin{bmatrix}
2 & -1 & 0 & 0 & \cdots & 0 \\
-1 & 2 & -1 & 0 & \cdots & 0 \\
0 & -1 & 2 & -1 & \cdots & 0 \\
\vdots & \cdots & & & \cdots & \vdots \\
0 & \cdots & 0 & -1 & 2 & -1 \\
0 & \cdots & 0 & 0 & -2 & 2
\end{bmatrix}
\begin{bmatrix}
u_2 \\ u_3 \\ u_4 \\ \vdots \\ u_{n-1} \\ u_n
\end{bmatrix}
=
\begin{bmatrix}
\Delta X p_0 \\
\Delta X p_0 \\
\Delta X p_0 \\
\vdots \\
\Delta X p_0 \\
\frac{\Delta X p_0}{2}
\end{bmatrix}
, \tag{6.7}
$$

where $\Delta X = \frac{L}{n-1}$ for equidistant spacing.
pg

Fig. 6.1 Finite difference
discretization of the
fixed-ended rod (single force
case) based on seven grid
nodes

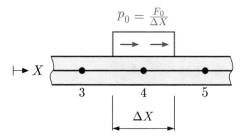

2.5 Displacement distribution for a fixed-ended rod structure

The finite difference discretization of the fixed-ended rod is shown in Fig. 6.1 for
seven domain nodes. The single force F_0 is understood as the integral value of a
constant distributed load p_0, which is acting over a length of ΔX.

The evaluation of the finite difference approximation according to Eq. (2.9) at the
inner nodes $i = 2, \ldots, 6$ gives:

$$\text{node 2:} \quad \frac{EA}{\Delta X}(u_3 - 2u_2 + u_1) = 0, \tag{6.8}$$

$$\text{node 3:} \quad \frac{EA}{\Delta X}(u_4 - 2u_3 + u_2) = 0, \tag{6.9}$$

$$\text{node 4:} \quad \frac{EA}{\Delta X}(u_5 - 2u_4 + u_3) = -p_0\Delta X = -F_0, \tag{6.10}$$

$$\text{node 5:} \quad \frac{EA}{\Delta X}(u_6 - 2u_5 + u_4) = 0, \tag{6.11}$$

$$\text{node 6:} \quad \frac{EA}{\Delta X}(u_7 - 2u_6 + u_5) = 0, \tag{6.12}$$

or in matrix notation under consideration of the boundary conditions:

$$\begin{bmatrix} -2 & 1 & 0 & 0 & 0 \\ 1 & -2 & 1 & 0 & 0 \\ 0 & 1 & -2 & 1 & 0 \\ 0 & 0 & 1 & -2 & 1 \\ 0 & 0 & 0 & 1 & -2 \end{bmatrix} \begin{bmatrix} u_2 \\ u_3 \\ u_4 \\ u_5 \\ u_6 \end{bmatrix} = -\frac{\Delta X F_0}{EA} \begin{bmatrix} 0 \\ 0 \\ 1 \\ 0 \\ 0 \end{bmatrix}. \tag{6.13}$$

The solution of this linear system of equations gives the unknown nodal values as:

$$\begin{bmatrix} u_2 \\ u_3 \\ u_4 \\ u_5 \\ u_6 \end{bmatrix} = \frac{F_0 L}{EA} \begin{bmatrix} \frac{1}{6} \\ \frac{1}{3} \\ \frac{1}{2} \\ \frac{1}{3} \\ \frac{1}{6} \end{bmatrix}. \tag{6.14}$$

Fig. 6.2 Finite difference discretization of the fixed-ended rod based on seven grid nodes: **a** axial point load; **b** load per length

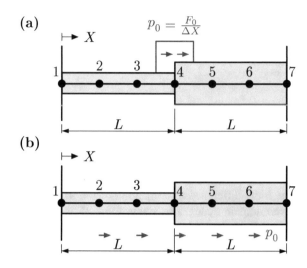

This result is identical to the analytical solution.

2.6 Displacement distribution for a fixed-ended stepped rod structure

The finite difference discretization of the fixed-ended rod is shown in Fig. 6.2 for seven domain nodes. The single force F_0 is understood as the integral value of a constant distributed load p_0, which is acting over a length of ΔX, see Fig. 6.2a.

(a) The evaluation of the finite difference approximation according to Eq. (2.56) at the inner nodes $i = 2, \ldots, 6$ gives:

$$\text{node 2:}\quad \frac{E A_{\mathrm{I}}}{\Delta X}\,(u_3 - 2u_2 + u_1) = 0\,, \tag{6.15}$$

$$\text{node 3:}\quad \frac{E A_{\mathrm{I}}}{\Delta X}\,(u_4 - 2u_3 + u_2) = 0\,, \tag{6.16}$$

$$\text{node 4:}\quad \frac{E}{\Delta X}\left(\frac{A_{\mathrm{II}} - A_{\mathrm{I}}}{4}\,(u_5 - u_3) + \frac{A_{\mathrm{I}} + A_{\mathrm{II}}}{2}\,(u_5 - 2u_4 + u_3)\right) =$$
$$- p_0\Delta X = -F_0\,, \tag{6.17}$$

$$\text{node 5:}\quad \frac{E A_{\mathrm{II}}}{\Delta X}\,(u_6 - 2u_5 + u_4) = 0\,, \tag{6.18}$$

$$\text{node 6:}\quad \frac{E A_{\mathrm{II}}}{\Delta X}\,(u_7 - 2u_6 + u_5) = 0\,, \tag{6.19}$$

or in matrix notation under consideration of the boundary conditions:

$$\begin{bmatrix} -2 & 1 & 0 & 0 & 0 \\ 1 & -2 & 1 & 0 & 0 \\ 0 & \frac{3A_I+A_{II}}{4} & -(A_I + A_{II}) & \frac{A_I+3A_{II}}{4} & 0 \\ 0 & 0 & 1 & -2 & 1 \\ 0 & 0 & 0 & 1 & -2 \end{bmatrix} \begin{bmatrix} u_2 \\ u_3 \\ u_4 \\ u_5 \\ u_6 \end{bmatrix} = -\frac{\Delta X F_0}{E} \begin{bmatrix} 0 \\ 0 \\ 1 \\ 0 \\ 0 \end{bmatrix}. \qquad (6.20)$$

The solution of this linear system of equations gives the unknown nodal values as:

$$\begin{bmatrix} u_2 \\ u_3 \\ u_4 \\ u_5 \\ u_6 \end{bmatrix} = \frac{F_0 L}{EA} \begin{bmatrix} \frac{1}{9} \\ \frac{2}{9} \\ \frac{1}{3} \\ \frac{2}{9} \\ \frac{1}{9} \end{bmatrix}. \qquad (6.21)$$

(a) The evaluation of the finite difference approximation according to Eq. (2.56) at the inner nodes $i = 2, \ldots, 6$ gives:

node 2: $\quad \dfrac{E A_I}{\Delta X} (u_3 - 2u_2 + u_1) = -p_0 \Delta X$, $\qquad\qquad\qquad$ (6.22)

node 3: $\quad \dfrac{E A_I}{\Delta X} (u_4 - 2u_3 + u_2) = -p_0 \Delta X$, $\qquad\qquad\qquad$ (6.23)

node 4: $\quad \dfrac{E}{\Delta X} \left(\dfrac{A_{II} - A_I}{4} (u_5 - u_3) + \dfrac{A_I + A_{II}}{2} (u_5 - 2u_4 + u_3) \right) =$

$$\qquad\qquad\qquad\qquad\qquad\qquad - p_0 \Delta X, \qquad (6.24)$$

node 5: $\quad \dfrac{E A_{II}}{\Delta X} (u_6 - 2u_5 + u_4) = -p_0 \Delta X$, $\qquad\qquad\qquad$ (6.25)

node 6: $\quad \dfrac{E A_{II}}{\Delta X} (u_7 - 2u_6 + u_5) = -p_0 \Delta X$, $\qquad\qquad\qquad$ (6.26)

or in matrix notation under consideration of the boundary conditions:

$$\begin{bmatrix} -2 & 1 & 0 & 0 & 0 \\ 1 & -2 & 1 & 0 & 0 \\ 0 & \frac{3A_I+A_{II}}{4} & -(A_I + A_{II}) & \frac{A_I+3A_{II}}{4} & 0 \\ 0 & 0 & 1 & -2 & 1 \\ 0 & 0 & 0 & 1 & -2 \end{bmatrix} \begin{bmatrix} u_2 \\ u_3 \\ u_4 \\ u_5 \\ u_6 \end{bmatrix} = -\frac{\Delta X^2 p_0}{E} \begin{bmatrix} 1/A_I \\ 1/A_I \\ 1 \\ 1/A_{II} \\ 1/A_{II} \end{bmatrix}. \qquad (6.27)$$

The solution of this linear system of equations gives the unknown nodal values as:

$$\begin{bmatrix} u_2 \\ u_3 \\ u_4 \\ u_5 \\ u_6 \end{bmatrix} = \frac{p_0 L^2}{EA} \begin{bmatrix} \frac{49}{216} \\ \frac{37}{108} \\ \frac{25}{72} \\ \frac{31}{108} \\ \frac{37}{216} \end{bmatrix}. \tag{6.28}$$

2.7 Elongation of a rod due to distributed load

The analytical solution can be taken from Ref. [1].
    Section $0 \le X \le a_1$:

$$u_\mathrm{I}(X) = \frac{p_0(a_2 - a_1)}{EA} \times X$$
$$= \frac{1}{3} \times \frac{p_0 L^2}{EA} \times \frac{X}{L}. \tag{6.29}$$

Section $a_1 \le X \le a_2$:

$$u_\mathrm{II}(X) = \frac{p_0}{EA} \left(-\frac{1}{2}(X - a_1)^2 + (a_2 - a_1)X \right)$$
$$= \frac{p_0 L^2}{EA} \left(-\frac{1}{2}\left(\frac{X}{L} - \frac{1}{3}\right)^2 + \frac{1}{3} \times \frac{X}{L} \right). \tag{6.30}$$

Section $a_2 \le X \le L$:

$$u_\mathrm{III}(X) = \frac{p_0}{EA} \left(-\frac{1}{2}\left(\frac{2L}{3} - a_1\right)^2 + (a_2 - a_1)\frac{2L}{3} \right)$$
$$= \frac{1}{6} \times \frac{p_0 L^2}{EA} = \mathrm{const}. \tag{6.31}$$

The finite difference discretization of the bi-material rod is shown in Fig. 6.3 for five domain nodes.

Fig. 6.3 Finite difference discretization of the rod loaded due to a distributed load in the segment $a_1 \le X \le a_2$

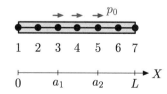

The evaluation of the finite difference approximation according to Eq. (2.9) at the nodes $i = 2, \ldots, 7$ gives:

$$\text{node 2:} \quad \frac{EA}{\Delta X} (u_3 - 2u_2 + u_1) = 0, \tag{6.32}$$

$$\text{node 3:} \quad \frac{EA}{\Delta X} (u_4 - 2u_3 + u_2) = -\frac{p_0 \Delta X}{2}, \tag{6.33}$$

$$\text{node 4:} \quad \frac{EA}{\Delta X} (u_5 - 2u_4 + u_3) = -p_0 \Delta X, \tag{6.34}$$

$$\text{node 5:} \quad \frac{EA}{\Delta X} (u_6 - 2u_5 + u_4) = -\frac{p_0 \Delta X}{2}, \tag{6.35}$$

$$\text{node 6:} \quad \frac{EA}{\Delta X} (u_7 - 2u_6 + u_5) = 0, \tag{6.36}$$

$$\text{node 7:} \quad \frac{EA}{\Delta X} (u_8 - 2u_7 + u_6) = 0, \tag{6.37}$$

or under consideration of the conditions as the boundaries, i.e. $u_1 = 0$ and $u_8 = u_6$,

$$\text{node 2:} \ -2u_2 + u_3 = 0, \tag{6.38}$$

$$\text{node 3:} \ u_2 - 2u_3 + u_4 = -\frac{p_0 \Delta X^2}{2EA}, \tag{6.39}$$

$$\text{node 4:} \ u_3 - 2u_4 + u_5 = -\frac{p_0 \Delta X^2}{EA}, \tag{6.40}$$

$$\text{node 5:} \ u_4 - 2u_5 + u_6 = -\frac{p_0 \Delta X^2}{2EA}, \tag{6.41}$$

$$\text{node 6:} \ u_5 - 2u_6 + u_7 = 0, \tag{6.42}$$

$$\text{node 7:} \ 2u_6 - 2u_7 = 0, \tag{6.43}$$

or in matrix form:

$$\begin{bmatrix} -2 & 1 & 0 & 0 & 0 & 0 \\ 1 & -2 & 1 & 0 & 0 & 0 \\ 0 & 1 & -2 & 1 & 0 & 0 \\ 0 & 0 & 1 & -2 & 1 & 0 \\ 0 & 0 & 0 & 1 & -2 & 1 \\ 0 & 0 & 0 & 0 & 2 & -2 \end{bmatrix} \begin{bmatrix} u_2 \\ u_3 \\ u_4 \\ u_5 \\ u_6 \\ u_7 \end{bmatrix} = -\frac{p_0 \Delta X^2}{EA} \begin{bmatrix} 0 \\ \frac{1}{2} \\ 1 \\ \frac{1}{2} \\ 0 \\ 0 \end{bmatrix}. \tag{6.44}$$

The solution of this linear system of equations gives the unknown nodal values as:

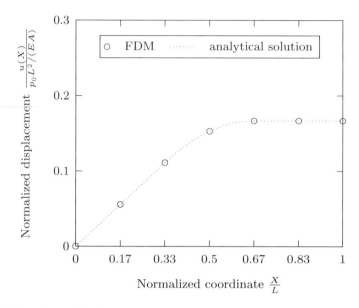

Fig. 6.4 Comparison of the displacements obtained from the finite difference approach and the exact analytical solution for the rod loaded due to a distributed load in the segment $a_1 \leq X \leq a_2$

$$
\begin{bmatrix} u_2 \\ u_3 \\ u_4 \\ u_5 \\ u_6 \\ u_7 \end{bmatrix} = \frac{p_0 L^2}{EA} \begin{bmatrix} \frac{1}{18} \\ \frac{1}{9} \\ \frac{11}{72} \\ \frac{1}{6} \\ \frac{1}{6} \\ \frac{1}{6} \end{bmatrix} = \frac{p_0 L^2}{EA} \begin{bmatrix} 0.055555556 \\ 0.11111111 \\ 0.15277778 \\ 0.16666667 \\ 0.16666667 \\ 0.16666667 \end{bmatrix} . \tag{6.45}
$$

The comparison between the FDM solution and the analytical solution is shown in Fig. 6.4. An excellent agreement between both solutions can be seen.

2.8 Elongation of a bi-material rod: finite difference solution and comparison with analytical solution

The analytical solution can be taken from Ref. [1].
    Section $0 \leq X \leq L$:

$$
u_{\mathrm{I}}(X) = \frac{1}{k_{\mathrm{I}}} \left(-\frac{p_0 x^2}{2} + \left[\frac{2k_{\mathrm{I}} + k_{\mathrm{II}}}{2(k_{\mathrm{I}} + k_{\mathrm{II}})} p_0 L + \frac{k_{\mathrm{I}} k_{\mathrm{II}}}{k_{\mathrm{I}} + k_{\mathrm{II}}} \frac{u_0}{L} \right] X \right) . \tag{6.46}
$$

Section $L \leq X \leq 2L$:

Fig. 6.5 Finite difference discretization of the bi-material rod

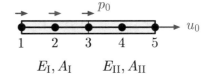

$$u_{\mathrm{II}}(X) = \frac{1}{k_{\mathrm{II}}} \left(\left[-\frac{k_{\mathrm{II}}}{2(k_{\mathrm{I}} + k_{\mathrm{II}})} p_0 L + \frac{k_{\mathrm{I}} k_{\mathrm{II}}}{k_{\mathrm{I}} + k_{\mathrm{II}}} \frac{u_0}{L} \right] X \right.$$
$$\left. + \frac{k_{\mathrm{II}}}{k_{\mathrm{I}} + k_{\mathrm{II}}} p_0 L^2 + \frac{k_{\mathrm{II}}(k_{\mathrm{II}} - k_{\mathrm{I}})}{k_{\mathrm{I}} + k_{\mathrm{II}}} u_0 \right). \tag{6.47}$$

Assigning the specific values $k_{\mathrm{I}} = 2k_{\mathrm{II}} = 1$, $L_{\mathrm{I}} = L_{\mathrm{II}} = 1$, $p_0 = 1$, and $u_0 = 1$, Eqs. (6.46) and (6.47) can be simplified to the following expression:

$$u_{\mathrm{I}}(X) = \frac{7X}{6} - \frac{X^2}{2}, \tag{6.48}$$

$$u_{\mathrm{II}}(X) = \frac{X}{3} + \frac{1}{3}. \tag{6.49}$$

The finite difference discretization of the bi-material rod is shown in Fig. 6.5 for five domain nodes.

The evaluation of the finite difference approximation according to Eq. (2.56) at the inner nodes $i = 2, \ldots, 5$ gives:

node 2: $\dfrac{1}{\Delta X} (0 + (EA)_{\mathrm{I}} (u_3 - 2u_2 + u_1)) = -p_0 \Delta X$, $\tag{6.50}$

node 3: $\dfrac{1}{\Delta X} \left(\dfrac{(EA)_{\mathrm{II}} - (EA)_{\mathrm{I}}}{4} (u_4 - u_2) + \dfrac{(EA)_{\mathrm{I}} + (EA)_{\mathrm{II}}}{2} (u_4 - 2u_3 + u_2) \right) =$
$$- p_0 \frac{\Delta X}{2} = -R, \tag{6.51}$$

node 4: $\dfrac{1}{\Delta X} (0 + (EA)_{\mathrm{II}} (u_5 - 2u_4 + u_3)) = 0$, $\tag{6.52}$

or in matrix notation under consideration of the boundary conditions, i.e., $u_1 = 0$ and $u_5 = u_0$:

$$\begin{bmatrix} -2 & 1 & 0 \\ \frac{3(EA)_{\mathrm{I}}+(EA)_{\mathrm{II}}}{4} & -((EA)_{\mathrm{I}} + (EA)_{\mathrm{II}}) & \frac{(EA)_{\mathrm{I}}+3(EA)_{\mathrm{II}}}{4} \\ 0 & 1 & -2 \end{bmatrix} \begin{bmatrix} u_2 \\ u_3 \\ u_4 \end{bmatrix} = \begin{bmatrix} -\frac{p_0 \Delta X^2}{(EA)_{\mathrm{I}}} \\ -\frac{p_0 \Delta X^2}{2} \\ -u_0 \end{bmatrix}. \tag{6.53}$$

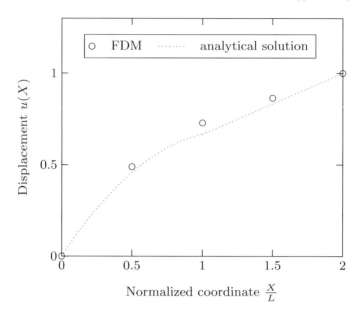

Fig. 6.6 Comparison of the displacements obtained from the finite difference approach and the exact analytical solution

The solution of this linear system of equations gives the unknown nodal values as:

$$
\begin{bmatrix} u_2 \\ u_3 \\ u_4 \end{bmatrix} = \begin{bmatrix} \frac{(3(EA)_{II}+(EA)_I)u_0}{4(2(EA)_{II}+2(EA)_I)} + \frac{L^2 p_0}{4(2(EA)_{II}+2(EA)_I)} + \frac{L^2 (5(EA)_{II}+7(EA)_I)p_0}{16(EA)_I(2(EA_{II}+2(EA)_I)} \\ \frac{(3(EA)_{II}+(EA)_I)u_0}{2(2(EA)_{II}+2(EA)_I)} + \frac{L^2 ((EA)_{II}+3(EA)_I)p_0}{8(EA)_I(2v_{II}+2(EA)_I)} + \frac{L^2 p_0}{2(2(EA)_{II}+2(EA)_I)} \\ \frac{(7(EA)_{II}+5(EA)_I)u_0}{4(2(EA)_{II}+2(EA)_I)} + \frac{L^2 ((EA)_{II}+3(EA)_I)p_0}{16(EA)_I(2(EA)_{II}+2(EA)_I)} + \frac{L^2 p_0}{4(2(EA)_{II}+2(EA)_I)} \end{bmatrix} , \quad (6.54)
$$

or under consideration of the specific values $k_I = 2k_{II} = 1$, $L_I = L_{II} = 1$, $p_0 = 1$, and $u_0 = 1$:

$$
\begin{bmatrix} u_2 \\ u_3 \\ u_4 \end{bmatrix} = \begin{bmatrix} \frac{47}{96} \\ \frac{35}{48} \\ \frac{83}{96} \end{bmatrix} = \begin{bmatrix} 0.48958333 \\ 0.72916667 \\ 0.86458333 \end{bmatrix} . \quad (6.55)
$$

The comparison between the FDM solution and the analytical solution is shown in Fig. 6.6. A quite good agreement between both solutions can be seen.

6.3   Answers for Problems from Chap. 3

3.7   Finite difference approximation of a simply supported and cantilevered beam based on three domain nodes

The single force F_0 is understood as the integral value of a distributed load q_0, which is acting over a length of (a) ΔX or (b) $\Delta X/2$.

(a) Simply supported beam:

$$\text{node 2:}\quad \frac{EI_Y}{\Delta X^3}(u_4 - 4u_3 + 6u_2 - 4u_1 + u_0) = -F_0. \tag{6.56}$$

Boundary conditions: $u(X = 0) = u_1 = 0$, $u(X=L)=u_3 = 0$, $M_Y(X = 0) = 0 \to u_0 = -u_2$, $M_Y(X = L) = 0 \to u_4 = -u_2$.

$$u_2 = -\frac{\Delta X^3 F_0}{4EI_Y} = -\underbrace{\frac{1}{32}}_{0.03125}\frac{F_0 L^3}{EI_Y}. \tag{6.57}$$

Relative error to analytical solution: 50%.

(b) Cantilevered beam:

$$\text{node 2:}\quad \frac{EI_Y}{\Delta X^3}(u_4 - 4u_3 + 6u_2 - 4u_1 + u_0) = 0, \tag{6.58}$$

$$\text{node 3:}\quad \frac{EI_Y}{\Delta X^3}(u_5 - 4u_4 + 6u_3 - 4u_2 + u_1) = -F_0. \tag{6.59}$$

Boundary conditions: $u(X = 0)=u_1=0$, $\frac{du_1}{dX} = 0 \to u_0 = u_2$, $EI_Y\frac{d^2u}{dX^2}\big|_3 = -M_Y = 0 \to u_4 = -u_2 + 2u_3$, $EI_Y\frac{d^3u}{dX^3}\big|_3 = 0 \to u_5 = 4u_3 - 4u_2 + 0$.

Linear system of equations:

$$\frac{EI_Y}{\Delta X^3}\begin{bmatrix} 3 & -1 \\ -4 & 2 \end{bmatrix}\begin{bmatrix} u_2 \\ u_3 \end{bmatrix} = \begin{bmatrix} 0 \\ -F_0 \end{bmatrix}. \tag{6.60}$$

Solution:

$$u_2 = -\frac{1}{2}\frac{\Delta X^3 F_0}{EI_Y} = -\frac{1}{16}\frac{F_0 L^3}{EI_Y}, \tag{6.61}$$

$$u_3 = -\frac{3}{2}\frac{\Delta X^3 F_0}{EI_Y} = -\underbrace{\frac{3}{16}}_{0.1875}\frac{F_0 L^3}{EI_Y}. \tag{6.62}$$

Relative error at node 3 compared to analytical solution: 43.75%.

3.8 Centered difference approximation of the fourth order derivative

$$u_{i+1} = u_i + \left(\frac{d^1 u}{dX^1}\right)_i \frac{\Delta X}{1!} + \left(\frac{d^2 u}{dX^2}\right)_i \frac{\Delta X^2}{2!} + \left(\frac{d^3 u}{dX^3}\right)_i \frac{\Delta X^3}{3!} +$$
$$+ \left(\frac{d^4 u}{dX^4}\right)_i \frac{\Delta X^4}{4!} + \left(\frac{d^5 u}{dX^5}\right)_i \frac{\Delta X^5}{5!} + \left(\frac{d^6 u}{dX^6}\right)_i \frac{\Delta X^6}{6!} + \cdots, \qquad (6.63)$$

$$u_{i-1} = u_i - \left(\frac{d^1 u}{dX^1}\right)_i \frac{\Delta X}{1!} + \left(\frac{d^2 u}{dX^2}\right)_i \frac{\Delta X^2}{2!} - \left(\frac{d^3 u}{dX^3}\right)_i \frac{\Delta X^3}{3!} +$$
$$+ \left(\frac{d^4 u}{dX^4}\right)_i \frac{\Delta X^4}{4!} - \left(\frac{d^5 u}{dX^5}\right)_i \frac{\Delta X^5}{5!} + \left(\frac{d^6 u}{dX^6}\right)_i \frac{\Delta X^6}{6!} - \cdots. \qquad (6.64)$$

Summing up the expressions (6.63) and (6.64) and rearranging for the second order derivative gives:

$$\left(\frac{d^2 u}{dX^2}\right)_i = \frac{u_{i+1} - 2u_1 + u_{i-1}}{\Delta X^2} - \left(\frac{d^4 u}{dX^4}\right)_i \frac{\Delta X^2}{12} - \left(\frac{d^6 u}{dX^6}\right)_i \frac{\Delta X^4}{360} - \cdots. \qquad (6.65)$$

$$u_{i+2} = u_i + \left(\frac{d^1 u}{dX^1}\right)_i \frac{(2\Delta X)}{1!} + \left(\frac{d^2 u}{dX^2}\right)_i \frac{(2\Delta X)^2}{2!} + \left(\frac{d^3 u}{dX^3}\right)_i \frac{(2\Delta X)^3}{3!} +$$
$$+ \left(\frac{d^4 u}{dX^4}\right)_i \frac{(2\Delta X)^4}{4!} + \left(\frac{d^5 u}{dX^5}\right)_i \frac{(2\Delta X)^5}{5!} + \left(\frac{d^6 u}{dX^6}\right)_i \frac{(2\Delta X)^6}{6!} + \cdots, \qquad (6.66)$$

$$u_{i-2} = u_i - \left(\frac{d^1 u}{dX^1}\right)_i \frac{(2\Delta X)}{1!} + \left(\frac{d^2 u}{dX^2}\right)_i \frac{(2\Delta X)^2}{2!} - \left(\frac{d^3 u}{dX^3}\right)_i \frac{(2\Delta X)^3}{3!} +$$
$$+ \left(\frac{d^4 u}{dX^4}\right)_i \frac{(2\Delta X)^4}{4!} - \left(\frac{d^5 u}{dX^5}\right)_i \frac{(2\Delta X)^5}{5!} + \left(\frac{d^6 u}{dX^6}\right)_i \frac{(2\Delta X)^6}{6!} - \cdots. \qquad (6.67)$$

Summing up the expressions (6.66) and (6.67) and considering the expression for the second order derivative given in Eq. (6.65) allows to express the fourth order derivative finally as:

$$\left(\frac{d^4 u}{dX^4}\right)_i = \frac{u_{i+2} - 4u_{i+1} + 6u_i - 4u_{i-1} + u_{i-2}}{(\Delta X)^4} - \underbrace{\left(\frac{d^6 u}{dX^6}\right)_i \frac{(\Delta X)^2}{6}}_{O(\Delta X^2)}. \qquad (6.68)$$

3.9 Finite difference approximation of a cantilevered beam based on five domain nodes

The single force F_0 is understood as the integral value of a distributed load q_0, which is acting over a length of ΔX.

Centered finite difference approximation at nodes $i = 2, \ldots, 5$:

$$\text{node 2:} \quad \frac{EI_Y}{\Delta X^3} (u_4 - 4u_3 + 6u_2 - 4u_1 + u_0) = 0, \tag{6.69}$$

$$\text{node 3:} \quad \frac{EI_Y}{\Delta X^3} (u_5 - 4u_4 + 6u_3 - 4u_2 + u_1) = 0, \tag{6.70}$$

$$\text{node 4:} \quad \frac{EI_Y}{\Delta X^3} (u_6 - 4u_5 + 6u_4 - 4u_3 + u_2) = -F_0, \tag{6.71}$$

$$\text{node 5:} \quad \frac{EI_Y}{\Delta X^3} (u_7 - 4u_6 + 6u_5 - 4u_4 + u_3) = 0. \tag{6.72}$$

Boundary conditions:

$$u_1 = 0, \quad \left.\frac{du}{dX}\right|_1 = 0 \;\rightarrow\; u_0 = u_2. \tag{6.73}$$

$$M_Y(L) = 0 \;\rightarrow\; u_6 = 2u_5 - u_4, \quad Q_Z(L) = 0 \;\rightarrow\; u_7 = 2u_6 - 2u_4 + u_3. \tag{6.74}$$

Linear system of equations:

$$\frac{EI_Y}{\Delta X^3}
\begin{bmatrix}
7 & -4 & 1 & 0 \\
-4 & 6 & -4 & 1 \\
1 & -4 & 5 & -2 \\
0 & 1 & -2 & 1
\end{bmatrix}
\begin{bmatrix}
u_2 \\
u_3 \\
u_4 \\
u_5
\end{bmatrix}
=
\begin{bmatrix}
0 \\
0 \\
-F_0 \\
0
\end{bmatrix}. \tag{6.75}$$

Solution:

$$u_2 = -\frac{3F_0L^3}{128EI_Y}, \quad u_3 = -\frac{5F_0L^3}{64EI_Y}, \quad u_4 = -\frac{19F_0L^3}{128EI_Y}, \quad u_5 = -\frac{7F_0L^3}{32EI_Y}. \tag{6.76}$$

The analytical solution can be extracted from [1] as $u(L) = -\frac{81F_0L^3}{384EI_Y}$ and the relative error is obtained as 3.704%.

3.10 Finite difference approximation of a cantilevered beam based on five domain nodes—backward scheme

(a)

$$\left. EI_Y \frac{d^3u}{dX^3} \right|_5 = -Q_Z(X = L) = 0. \tag{6.77}$$

$$\frac{E I_Y}{\Delta X^3} \begin{bmatrix} 7 & -4 & 1 & 0 \\ -4 & 6 & -4 & 1 \\ 1 & -4 & 5 & -2 \\ -14 & 24 & -18 & 5 \end{bmatrix} \begin{bmatrix} u_2 \\ u_3 \\ u_4 \\ u_5 \end{bmatrix} = \begin{bmatrix} 0 \\ 0 \\ -F_0 \\ 0 \end{bmatrix} . \tag{6.78}$$

$$\text{relative error} = \left| \frac{\frac{1}{8} + \frac{81}{384}}{-\frac{81}{384}} \right| \times 100 = 159.259\% . \tag{6.79}$$

(b)

$$E I_Y \left. \frac{d^3 u}{dX^3} \right|_4 = -Q_Z \left(X = \frac{3L}{4} \right) = F_0 . \tag{6.80}$$

$$\frac{E I_Y}{\Delta X^3} \begin{bmatrix} 7 & -4 & 1 & 0 \\ -4 & 6 & -4 & 1 \\ 1 & -4 & 5 & -2 \\ 27 & -18 & 5 & 0 \end{bmatrix} \begin{bmatrix} u_2 \\ u_3 \\ u_4 \\ u_5 \end{bmatrix} = \begin{bmatrix} 0 \\ 0 \\ -F_0 \\ 2F_0 \end{bmatrix} . \tag{6.81}$$

$$\text{relative error} = \frac{\frac{7}{32} - \frac{81}{384}}{\frac{81}{384}} \times 100 = 3.704\% . \tag{6.82}$$

3.11 Finite difference approximation of a cantilevered beam based on five domain nodes—conversion of tip load into distributed load

The final system of equations is obtained for this case as:

$$\frac{E I_Y}{\Delta X^3} \begin{bmatrix} 7 & -4 & 1 & 0 \\ -4 & 6 & -4 & 1 \\ 1 & -4 & 5 & -2 \\ 0 & 2 & -4 & 2 \end{bmatrix} \begin{bmatrix} u_2 \\ u_3 \\ u_4 \\ u_5 \end{bmatrix} = \begin{bmatrix} 0 \\ 0 \\ -q_0 \frac{\Delta X}{2} \\ -q_0 \frac{\Delta X}{2} \end{bmatrix} , \tag{6.83}$$

from which we get $u_5 = -\frac{25}{128} \frac{F_0 L^3}{E I_Y}$ and a relative error of 41.406%.

3.12 Finite difference approximation of a simply supported beam based on five domain nodes—bending moment approach

The bending moment distribution is obtained as $M_Y(X) = -\frac{q_0 X}{2}(L - X)$. Thus, the finite difference approximation is given by:

$$E I_Y \times \frac{u_{i+1} - 2u_i + u_{i-1}}{\Delta X^2} = \frac{q_0 X_i}{2} (L - X_i) . \tag{6.84}$$

Evaluation for nodes 2, 3 and 4 and consideration of the boundary conditions, i.e. $u_1 = u_5 = 0$, gives:

$$
\begin{bmatrix} -2 & 1 & 0 \\ 1 & -2 & 1 \\ 0 & 1 & -2 \end{bmatrix} \begin{bmatrix} u_2 \\ u_3 \\ u_4 \end{bmatrix} = \begin{bmatrix} \frac{q_0 \Delta X^3}{2EI_Y} (L - \Delta X) \\ \frac{q_0 \Delta X^3}{EI_Y} (L - 2\Delta X) \\ \frac{3q_0 \Delta X^3}{2EI_Y} (L - 3\Delta X) \end{bmatrix} . \tag{6.85}
$$

Solution:

$$
u_2 = -\frac{5q_0 L^4}{512EI_Y} , \quad u_3 = -\frac{7q_0 L^4}{512EI_Y} , \quad u_4 = -\frac{5q_0 L^4}{512EI_Y} . \tag{6.86}
$$

3.13   Finite difference approximation of a simply supported beam under pure bending

The finite difference approximations of the the 2nd and 4th order partial differential equation can be written as:

$$
EI_Y \left(\frac{u_{i+1} - 2u_i + u_{i-1}}{\Delta X^2} \right) = -M_i , \tag{6.87}
$$

$$
\frac{EI_Y}{\Delta X^3} (u_{i+2} - 4u_{i+1} + 6u_i - 4u_{i-1} + u_{i-2}) = R_i = 0 . \tag{6.88}
$$

(a) Five domain nodes
The system of linear equation based on the 2nd order PDE is given by:

$$
\frac{EI_Y}{\Delta X^2} \begin{bmatrix} -2 & 1 & 0 \\ 1 & -2 & 1 \\ 0 & 1 & -2 \end{bmatrix} \begin{bmatrix} u_2 \\ u_3 \\ u_4 \end{bmatrix} = \begin{bmatrix} M_0 \\ M_0 \\ M_0 \end{bmatrix} , \tag{6.89}
$$

from which its solution is obtained as:

$$
u_2 = u_4 = -\frac{3M_0 L^2}{32EI_Y} , \quad u_3 = -\frac{M_0 L^2}{8EI_Y} . \tag{6.90}
$$

The system of linear equation based on the 4th order PDE is given by:

$$
\frac{EI_Z}{\Delta X^3} \begin{bmatrix} 5 & -4 & 1 \\ -4 & 6 & -4 \\ 1 & -4 & 5 \end{bmatrix} \begin{bmatrix} u_2 \\ u_3 \\ u_4 \end{bmatrix} = \begin{bmatrix} -M_0/\Delta X \\ 0 \\ -M_0/\Delta X \end{bmatrix} , \tag{6.91}
$$

which gives the same results as in the case of the 2nd order PDE. Furthermore, the finite difference and analytical solutions are identical, cf. [1].
(b) Ten domain nodes
The system of linear equation based on the 2nd order PDE is given by:

$$\frac{EI_Y}{\Delta X^2}\begin{bmatrix} -2 & 1 & 0 & 0 & 0 & 0 & 0 & 0 & 0 \\ 1 & -2 & 1 & 0 & 0 & 0 & 0 & 0 & 0 \\ 0 & 1 & -2 & 1 & 0 & 0 & 0 & 0 & 0 \\ 0 & 0 & 1 & -2 & 1 & 0 & 0 & 0 & 0 \\ 0 & 0 & 0 & 1 & -2 & 1 & 0 & 0 & 0 \\ 0 & 0 & 0 & 0 & 1 & -2 & 1 & 0 & 0 \\ 0 & 0 & 0 & 0 & 0 & 1 & -2 & 1 & 0 \\ 0 & 0 & 0 & 0 & 0 & 0 & 1 & -2 & 1 \\ 0 & 0 & 0 & 0 & 0 & 0 & 0 & 1 & -2 \end{bmatrix}\begin{bmatrix} u_2 \\ u_3 \\ u_4 \\ u_5 \\ u_6 \\ u_7 \\ u_8 \\ u_9 \\ u_{10} \end{bmatrix}=\begin{bmatrix} M_0 \\ M_0 \\ M_0 \\ M_0 \\ M_0 \\ M_0 \\ M_0 \\ M_0 \\ M_0 \end{bmatrix}, \tag{6.92}$$

from which its solution is obtained as:

$$u_2 = -\frac{9M_0L^2}{200EI_Y} \quad, \quad u_3 = -\frac{2M_0L^2}{25EI_Y} \quad, \quad u_4 = -\frac{21M_0L^2}{200EI_Y}, \tag{6.93}$$

$$u_5 = -\frac{3M_0L^2}{25EI_Y} \quad, \quad u_6 = -\frac{M_0L^2}{8EI_Y}. \tag{6.94}$$

The system of linear equation based on the 4th order PDE is given by:

$$\frac{EI_Y}{\Delta X^3}\begin{bmatrix} 5 & -4 & 1 & 0 & 0 & 0 & 0 & 0 & 0 \\ -4 & 6 & -4 & 1 & 0 & 0 & 0 & 0 & 0 \\ 1 & -4 & 6 & -4 & 1 & 0 & 0 & 0 & 0 \\ 0 & 1 & -4 & 6 & -4 & 1 & 0 & 0 & 0 \\ 0 & 0 & 1 & -4 & 6 & -4 & 1 & 0 & 0 \\ 0 & 0 & 0 & 1 & -4 & 6 & -4 & 1 & 0 \\ 0 & 0 & 0 & 0 & 1 & -4 & 6 & -4 & 1 \\ 0 & 0 & 0 & 0 & 0 & 1 & -4 & 6 & -4 \\ 0 & 0 & 0 & 0 & 0 & 0 & 1 & -4 & 5 \end{bmatrix}\begin{bmatrix} u_2 \\ u_3 \\ u_4 \\ u_5 \\ u_6 \\ u_7 \\ u_8 \\ u_9 \\ u_{10} \end{bmatrix}=\begin{bmatrix} -M_0/\Delta X \\ 0 \\ 0 \\ 0 \\ 0 \\ 0 \\ 0 \\ 0 \\ -M_0/\Delta X \end{bmatrix}, \tag{6.95}$$

which gives the same results as in the case of the 2nd order PDE. Furthermore, the finite difference and analytical solutions are again identical, cf. [1].

3.14 Finite difference approximation of a simply supported beam—displacement, bending moment and shear force distribution

The single force F_0 is understood as the integral value of a distributed load q_0, which is acting over a length of ΔX.

The FD solution for the displacement can be taken from Example 3.1a as:

$$u_1 = u_5 = 0 \, , u_2 = u_4 = \frac{F_0L^3}{64EI_Y} \, , u_3 = \frac{3F_0L^3}{128EI_Y}. \tag{6.96}$$

FD moment distribution:

$$EI_Y \frac{d^2u}{dX^2} = -M_Y \quad \rightarrow \quad (EI_Y)_i \frac{u_{i+1} - 2u_i + u_{i-1}}{\Delta X^2} = -M_{Z,i} . \tag{6.97}$$

$$\rightarrow M_{Y,2} = M_{Y,4} = +\frac{F_0 L}{8} , M_{Y,3} = +\frac{F_0 L}{4} . \tag{6.98}$$

The boundary values, i.e. $M_{Y,1} = M_{Y,5} = 0$ result from the BC or from a forward ($i = 1$) or backward ($i = 5$) scheme.
FD shear force distribution:

$$EI_Y \frac{d^3u}{dX^3} = -Q_Z \quad \rightarrow \quad (EI_Y)_i \frac{u_{i+2} - 2u_{i+1} + 2u_{i-1} - u_{i-2}}{2\Delta X^3} = -Q_{Z,i} . \tag{6.99}$$

$$Q_{Z,3} = -\frac{64EI_Y}{2L^3} \big(\underbrace{u_5}_{0} - 2u_4 + 2u_2 - \underbrace{u_1}_{0} \big) = 0 . \tag{6.100}$$

The boundary values, i.e. $Q_{Z,1} = 2F_0$ and $Q_{Z,5} = -2F_0$ result from a forward ($i = 1$) or backward ($i = 5$) scheme, respectively. The values $Q_{Z,3}$ and $Q_{Z,4}$ cannot be calculated based on the actual subdivision.
Analytical solution ($0 \le X \le \frac{L}{2}$):

$$u(X) = \frac{F_0}{48EI_Y} \left(3L^2 X - 4X^3 \right) , M_Y(X) = +\frac{F_0 X}{2} , Q_y(X) = +\frac{F_0}{2} . \tag{6.101}$$

3.15 Finite difference approximation of a simply supported beam on elastic foundation based on five domain nodes

The single force F_0 is understood as the integral value of a distributed load q_0, which is acting over a length of ΔX.

Finite difference approximation of the partial differential equation:

$$EI_Y \times \frac{u_{i+2} - 4u_{i+1} + 6u_i - 4u_{i-1} + u_{i-2}}{\Delta X^3} + ku_i \Delta X = R_i . \tag{6.102}$$

Linear system of equations:

$$\begin{bmatrix} 5 + \frac{k\Delta X^4}{EI_Y} & -4 & 1 \\ -4 & 6 + \frac{k\Delta X^4}{EI_Y} & -4 \\ 1 & -4 & 5 + \frac{k\Delta X^4}{EI_Y} \end{bmatrix} \begin{bmatrix} u_2 \\ u_3 \\ u_4 \end{bmatrix} = \begin{bmatrix} 0 \\ -\frac{F_0\Delta X^3}{EI_Y} \\ 0 \end{bmatrix} . \tag{6.103}$$

Solution:

$$u_3 = u \left(\frac{L}{2} \right) = -\frac{\left(6EI_Y + \frac{kL^4}{256} \right) L^3 F_0}{64 \left(4(EI_Y)^2 + \frac{3EI_Y kL^4}{64} + \frac{k^2 L^8}{65536} \right)} . \tag{6.104}$$

Special case $k = 4$, $E I_Y = 1$ and $L = 1$:

$$u_3 = u\left(\frac{L}{2}\right) = -0.0224451 F_0 . \tag{6.105}$$

Relative error (analytical solution cf. [1]):

$$\text{relative error} = \frac{-0.0224451 F_0 + 0.0200233 F_0}{-0.0200233 F_0} \times 100 = 12.094\% . \tag{6.106}$$

3.16 Finite difference approximation of a simply supported beam with varying bending stiffness—constant distributed load

With the bending stiffness and bending moment distribution, the following differential equation can be written:

$$\frac{E I_0}{1 + \left(\frac{2X}{L} - 1\right)^2} \times \frac{d^2 u_Z(X)}{dX^2} = \frac{q_0 X}{2} (L - X) . \tag{6.107}$$

The following finite difference scheme can be derived:

$$u_{i+1} - 2u_i + u_{i-1} = \Delta X^2 \times \frac{q_0 X_i}{2} (L - X_i) \times \frac{1 + \left(\frac{2X_i}{L} - 1\right)^2}{E I_0} . \tag{6.108}$$

Under consideration of the boundary conditions, i.e. $u_1 = u_5 = 0$, the following system of equations is obtained:

$$\begin{bmatrix} -2 & 1 & 0 \\ 1 & -2 & 1 \\ 0 & 1 & -2 \end{bmatrix} \begin{bmatrix} u_2 \\ u_3 \\ u_4 \end{bmatrix} = \frac{q_0 L^4}{E I_0} \begin{bmatrix} \frac{15}{2048} \\ \frac{1}{128} \\ \frac{15}{2048} \end{bmatrix} . \tag{6.109}$$

Solution:

$$u_2 = -\frac{23 q_0 L^4}{2048 E I_0} , \quad u_3 = -\frac{31 q_0 L^4}{2048 E I_0} , \quad u_4 = -\frac{23 q_0 L^4}{2048 E I_0} . \tag{6.110}$$

Analytical solution:

$$u_Z(X) = -\frac{q_0}{E I_0 L^2} \left(\frac{1}{15} X^6 - \frac{1}{12} X^4 L^2 - \frac{1}{6} X^3 L^3\right) + c_1 X + c_2 . \tag{6.111}$$

Consideration of the boundary conditions gives $c_2 = 0$ and $c_1 = -\frac{11 q_0 L^3}{60 E I_0}$. Based on the analytical solution, the displacement in the middle of the beam is obtained as:

$$u_Z\left(\frac{L}{2}\right) = -\frac{q_0 L^4}{15 E I_0}.$$ (6.112)

$$\text{relative error} = \frac{-\frac{31}{2048} + \frac{1}{15}}{-\frac{1}{15}} \times 100 = -77.295\%.$$ (6.113)

3.17 Finite difference approximation of a simply supported beam with varying bending stiffness—constant bending moment

With the bending stiffness and bending moment distribution, the following differential equation can be written:

$$\frac{E I_0}{1 + \left(\frac{2X}{L} - 1\right)^2} \times \frac{\mathrm{d}^2 u_Z(X)}{\mathrm{d} X^2} = M_0.$$ (6.114)

The following finite difference scheme can be derived:

$$u_{i+1} - 2u_i + u_{i-1} = \Delta X^2 \times M_0 \times \frac{1 + \left(\frac{2X_i}{L} - 1\right)^2}{E I_0}.$$ (6.115)

Under consideration of the boundary conditions, i.e. $u_1 = u_5 = 0$, the following system of equations is obtained:

$$\begin{bmatrix} -2 & 1 & 0 \\ 1 & -2 & 1 \\ 0 & 1 & -2 \end{bmatrix} \begin{bmatrix} u_2 \\ u_3 \\ u_4 \end{bmatrix} = \frac{M_0 L^2}{E I_0} \begin{bmatrix} \frac{5}{64} \\ \frac{1}{16} \\ \frac{5}{64} \end{bmatrix}.$$ (6.116)

Solution:

$$u_2 = -\frac{7 M_0 L^2}{64 E I_0}, \quad u_3 = -\frac{9 M_0 L^2}{64 E I_0}, \quad u_4 = -\frac{7 M_0 L^2}{64 E I_0}.$$ (6.117)

Analytical solution:

$$u_Z(X) = \frac{2 M_0}{E I_0 L^2}\left(\frac{1}{2} X^2 L^2 - \frac{1}{3} X^3 L + \frac{1}{6} X^4\right) + c_1 X + C_2.$$ (6.118)

Consideration of the boundary conditions gives $c_2 = 0$ and $c_1 = -\frac{2 M_0 L}{E I_0}$. Based on the analytical solution, the displacement in the middle of the beam is obtained as:

$$u_Z\left(\frac{L}{2}\right) = -\frac{31 M_0 L^2}{48 E I_0}.$$ (6.119)

$$\text{relative error} = \frac{-\frac{9}{64} + \frac{31}{48}}{-\frac{31}{48}} \times 100 = -78.226\%.$$ (6.120)

Table 6.1 Values of the distributed load at the grid points

Grid point	Coordinate X	$q_Z(X)$
1	0	$-\alpha q_0$
2	$\frac{L}{4}$	$-\left(\frac{3}{4}\alpha + \frac{1}{4}\beta\right)q_0$
3	$\frac{L}{2}$	$-\left(\frac{1}{2}\alpha + \frac{1}{2}\beta\right)q_0$
4	$\frac{3L}{4}$	$-\left(\frac{1}{4}\alpha + \frac{3}{4}\beta\right)q_0$
5	L	$-\beta q_0$

3.18 Finite difference approximation of a simply supported beam loaded by a linearly distributed load

The function of the distributed load can be expressed as

$$q_Z(X) = -q_0\left(\alpha + (\beta - \alpha)\frac{X}{L}\right), \tag{6.121}$$

whereas the values at the five grid points are collected in Table 6.1.

Evaluation of the finite difference approximation of the fourth-order differential equation according to Eq. (3.9) at the inner nodes $i = 2, \ldots, 4$ and consideration of the boundary conditions gives:

$$\text{node 2:} \quad \frac{EI_Y}{\Delta X^3}(5u_2 - 4u_3 + u_4) = -\Delta X\left(\frac{3}{4}\alpha + \frac{1}{4}\beta\right)q_0, \tag{6.122}$$

$$\text{node 3:} \quad \frac{EI_Y}{\Delta X^3}(-4u_2 + 6u_3 - 4u_4) = -\Delta X\left(\frac{1}{2}\alpha + \frac{1}{2}\beta\right)q_0, \tag{6.123}$$

$$\text{node 4:} \quad \frac{EI_Y}{\Delta X^3}(u_2 - 4u_3 + 5u_4) = -\Delta X\left(\frac{1}{4}\alpha + \frac{3}{4}\beta\right)q_0. \tag{6.124}$$

or in matrix notation with $\Delta X = L/4$:

$$\begin{bmatrix} 5 & -4 & 1 \\ -4 & 6 & -4 \\ 1 & -4 & 5 \end{bmatrix} \begin{bmatrix} u_2 \\ u_3 \\ u_4 \end{bmatrix} = -\frac{q_0 L^4}{256 E I_Y} \begin{bmatrix} \left(\frac{3}{4}\alpha + \frac{1}{4}\beta\right) \\ \left(\frac{1}{2}\alpha + \frac{1}{2}\beta\right) \\ \left(\frac{1}{4}\alpha + \frac{3}{4}\beta\right) \end{bmatrix}. \tag{6.125}$$

The solution of this linear system of equations gives the unknown nodal values as:

$$\begin{bmatrix} u_2 \\ u_3 \\ u_4 \end{bmatrix} = -\frac{q_0 L^4}{4096 E I_Y} \begin{bmatrix} 21\alpha + 19\beta \\ 28\alpha + 28\beta \\ 19\alpha + 21\beta \end{bmatrix}. \tag{6.126}$$

Table 6.2 Values of the inner bending moment at the grid points

Grid point	Coordinate X	$M_Y(X)$
1	0	$\frac{3F_0L}{2}$
2	$\frac{L}{4}$	F_0L
3	$\frac{L}{2}$	$\frac{F_0L}{2}$
4	$\frac{3L}{4}$	$\frac{F_0L}{4}$
5	L	0

3.19 Finite difference approximation of a stepped cantilevered Euler–Bernoulli beam with two single forces based on five domain nodes

The internal bending moment distribution is obtained as

$$M_y(X) = F_0 \left(\frac{L}{2} - X \right) + F_0 (L - X) \quad \text{for } 0 \le X \le \frac{L}{2}, \tag{6.127}$$

$$M_Y(X) = F_0(L - X) \quad \text{for } \frac{L}{2} \le X \le L, \tag{6.128}$$

whereas the values at the five grid points are collected in Table 6.2.

Equation (3.27) must be evaluated for four different grid points since we have four unknown displacements $(u_2, \ldots, u_5)$ to determine. Considering grid points $i = 2, \ldots, 5$ introduces the fictitious node 6 at the right-hand side. However, this node cannot be eliminated based on the moment relation since we used this relationship already for node 5. Furthermore, the shear force relation would introduce a further fictitious node. Thus, it is recommended to state Eq. (3.27) for the grid points $i = 1, \ldots, 4$:

$$\text{node 1: } \frac{E2I_Y}{\Delta X^2} (u_2 - 2u_1 + u_0) = -\frac{3F_0L}{2}, \tag{6.129}$$

$$\text{node 2: } \frac{E2I_Y}{\Delta X^2} (u_3 - 2u_2 + u_1) = -F_0L, \tag{6.130}$$

$$\text{node 3: } \frac{E \left(\frac{2I_Y + I_Y}{2} \right)}{\Delta X^2} (u_4 - 2u_3 + u_2) = -\frac{F_0L}{2}, \tag{6.131}$$

$$\text{node 4: } \frac{EI_Y}{\Delta X^2} (u_5 - 2u_4 + u_3) = -\frac{F_0L}{4}, \tag{6.132}$$

or in matrix notation under consideration of the conditions at the left-hand boundary, i.e., $u_1 = 0$ and $u_0 = u_2$:

$$
\begin{bmatrix} 2 & 0 & 0 & 0 \\ -2 & 1 & 0 & 0 \\ 1 & -2 & 1 & 0 \\ 0 & 1 & -2 & 1 \end{bmatrix} \begin{bmatrix} u_2 \\ u_3 \\ u_4 \\ u_5 \end{bmatrix} = -\frac{\Delta X^2 F_0 L}{E I_Y} \begin{bmatrix} \frac{3}{4} \\ \frac{1}{2} \\ \frac{1}{3} \\ \frac{1}{4} \end{bmatrix}.
\tag{6.133}
$$

The solution of this linear system of equations gives the unknown nodal values as:

$$
\begin{bmatrix} u_2 \\ u_3 \\ u_4 \\ u_5 \end{bmatrix} = -\frac{F_0 L^3}{E I_Y} \begin{bmatrix} \frac{3}{128} \\ \frac{5}{64} \\ \frac{59}{384} \\ \frac{47}{192} \end{bmatrix} = -\frac{F_0 L^3}{E I_Y} \begin{bmatrix} 0.023438 \\ 0.078125 \\ 0.153646 \\ 0.244792 \end{bmatrix}.
\tag{6.134}
$$

3.20   Finite difference approximation of a fixed-ended beam with a distributed load

The finite difference discretization of the fixed-ended beam is shown in Fig. 6.7 for five and nine domain nodes.

(a) Evaluation of the finite difference approximation of the fourth-order differential equation according to Eq. (3.9) at the inner nodes $i = 2, \ldots, 4$ gives:

$$
\text{node 2: } \frac{E I_Y}{\Delta X^3} (u_4 - 4u_3 + 6u_2 - 4u_1 + u_0) = -q_0 \Delta X,
\tag{6.135}
$$

$$
\text{node 3: } \frac{E I_Y}{\Delta X^3} (u_5 - 4u_4 + 6u_3 - 4u_2 + u_1) = -q_0 \Delta X,
\tag{6.136}
$$

$$
\text{node 4: } \frac{E I_Y}{\Delta X^3} (u_6 - 4u_5 + 6u_4 - 4u_3 + u_2) = -q_0 \Delta X,
\tag{6.137}
$$

or under consideration of the boundary conditions, i.e. $u_1 = u_5 = 0$, $u_0 = u_2$ and $u_6 = u_4$, in matrix notation:

$$
\begin{bmatrix} 7 & -4 & 1 \\ -4 & 6 & -4 \\ 1 & -4 & 7 \end{bmatrix} \begin{bmatrix} u_2 \\ u_3 \\ u_4 \end{bmatrix} = -\frac{q_0 \Delta X^4}{E I_Y} \begin{bmatrix} 1 \\ 1 \\ 1 \end{bmatrix}.
\tag{6.138}
$$

The solution of this linear system of equations gives the unknown nodal values as:

$$
\begin{bmatrix} u_2 \\ u_3 \\ u_4 \end{bmatrix} = -\frac{q_0 L^4}{E I_Y} \begin{bmatrix} \frac{5}{2048} \\ \frac{1}{256} \\ \frac{5}{2048} \end{bmatrix},
\tag{6.139}
$$

and the relative error in the middle of the beam is obtained as [1]:

(a)

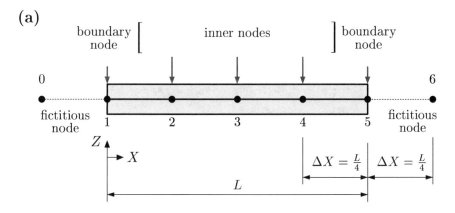

(b)

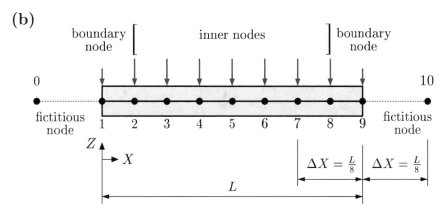

Fig. 6.7 Finite difference discretization of the fixed-ended Euler–Bernoulli beam (distributed load) based on **a** five and **b** nine grid nodes

$$\text{relative error} = \left| \frac{\frac{1}{256} - \frac{1}{384}}{\frac{1}{384}} \right| \times 100 = 50.0\% . \tag{6.140}$$

(b) Evaluation of the finite difference approximation of the fourth-order differential equation according to Eq. (3.9) at the inner nodes $i = 2, \ldots, 8$ gives:

$$\text{node 2:} \quad \frac{EI_Y}{\Delta X^3} (u_4 - 4u_3 + 6u_2 - 4u_1 + u_0) = -q_0 \Delta X , \tag{6.141}$$

$$\text{node 3:} \quad \frac{EI_Y}{\Delta X^3} (u_5 - 4u_4 + 6u_3 - 4u_2 + u_1) = -q_0 \Delta X , \tag{6.142}$$

$$\text{node 4:} \quad \frac{EI_Y}{\Delta X^3} (u_6 - 4u_5 + 6u_4 - 4u_3 + u_2) = -q_0 \Delta X , \tag{6.143}$$

node 5: $\dfrac{EI_Y}{\Delta X^3}(u_7 - 4u_6 + 6u_5 - 4u_4 + u_3) = -q_0\Delta X$,   (6.144)

node 6: $\dfrac{EI_Y}{\Delta X^3}(u_8 - 4u_7 + 6u_6 - 4u_5 + u_4) = -q_0\Delta X$,   (6.145)

node 7: $\dfrac{EI_Y}{\Delta X^3}(u_9 - 4u_8 + 6u_7 - 4u_6 + u_5) = -q_0\Delta X$,   (6.146)

node 8: $\dfrac{EI_Y}{\Delta X^3}(u_{10} - 4u_9 + 6u_8 - 4u_7 + u_6) = -q_0\Delta X$ (6.147)

or under consideration of the boundary conditions, i.e. $u_1 = u_9 = 0$, $u_0 = u_2$ and $u_{10} = u_8$, in matrix notation:

$$
\begin{bmatrix}
7 & -4 & 1 & 0 & 0 & 0 & 0 \\
-4 & 6 & -4 & 1 & 0 & 0 & 0 \\
1 & -4 & 6 & -4 & 1 & 0 & 0 \\
0 & 1 & -4 & 6 & -4 & 1 & 0 \\
0 & 0 & 1 & -4 & 6 & -4 & 1 \\
0 & 0 & 0 & 1 & -4 & 6 & -4 \\
0 & 0 & 0 & 0 & 1 & -4 & 7
\end{bmatrix}
\begin{bmatrix}
u_2 \\ u_3 \\ u_4 \\ u_5 \\ u_6 \\ u_7 \\ u_8
\end{bmatrix}
= -\frac{q_0\Delta X^4}{EI_Y}
\begin{bmatrix}
1 \\ 1 \\ 1 \\ 1 \\ 1 \\ 1 \\ 1
\end{bmatrix}.
\tag{6.148}
$$

The solution of this linear system of equations gives the unknown nodal values as:

$$
\begin{bmatrix}
u_2 \\ u_3 \\ u_4 \\ u_5 \\ u_6 \\ u_7 \\ u_8
\end{bmatrix}
= -\frac{q_0 L^4}{EI_Y}
\begin{bmatrix}
\frac{21}{32768} \\[4pt]
\frac{7}{4096} \\[4pt]
\frac{85}{32768} \\[4pt]
\frac{3}{1024} \\[4pt]
\frac{85}{32768} \\[4pt]
\frac{7}{4096} \\[4pt]
\frac{21}{32768}
\end{bmatrix},
\tag{6.149}
$$

and the relative error in the middle of the beam is obtained as [1]:

$$
\text{relative error} = \left| \frac{\frac{3}{1024} - \frac{1}{384}}{\frac{1}{384}} \right| \times 100 = 12.5\%.
\tag{6.150}
$$

3.21 Finite difference approximation of a fixed-ended beam with a single load

The finite difference discretization of the fixed-ended beam is shown in Fig. 6.8 for five and nine domain nodes.

The solution approach is based on the idea that the single force F_0 can be modeled as a distributed load q_0, which acts over a length of ΔX, see Fig. 6.9.

(a)

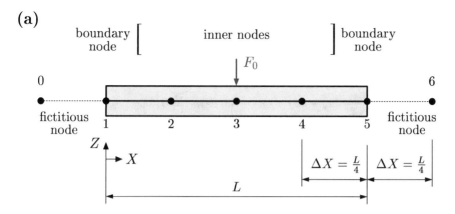

(b)

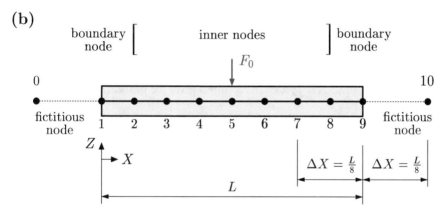

Fig. 6.8 Finite difference discretization of the fixed-ended Euler–Bernoulli beam (single force) based on **a** five and **b** nine grid nodes

Fig. 6.9 Modeling approach to consider a single force F_0 as a distributed load q_0 in the fourth-order differential equation

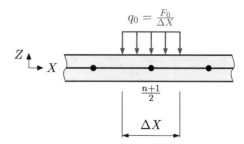

Thus, we can apply the finite difference approximation of the fourth-order differential equation according to Eq. (3.9). Writing this statement at the inner nodes $i = 2, \ldots, 4$ gives:

node 2: $\dfrac{E I_Y}{\Delta X^3} (u_4 - 4u_3 + 6u_2 - 4u_1 + u_0) = 0$, $\qquad$ (6.151)

node 3: $\dfrac{E I_Y}{\Delta X^3} (u_5 - 4u_4 + 6u_3 - 4u_2 + u_1) = -q_0 \Delta X = -F_0$, $\qquad$ (6.152)

node 4: $\dfrac{E I_Y}{\Delta X^3} (u_6 - 4u_5 + 6u_4 - 4u_3 + u_2) = 0$, $\qquad$ (6.153)

or under consideration of the boundary conditions, i.e. $u_1 = u_5 = 0$, $u_0 = u_2$ and $u_6 = u_4$, in matrix notation:

$$
\begin{bmatrix} 7 & -4 & 1 \\ -4 & 6 & -4 \\ 1 & -4 & 7 \end{bmatrix}
\begin{bmatrix} u_2 \\ u_3 \\ u_4 \end{bmatrix}
= -\frac{F_0 \Delta X^3}{E I_Y}
\begin{bmatrix} 0 \\ 1 \\ 0 \end{bmatrix} .
\qquad (6.154)
$$

The solution of this linear system of equations gives the unknown nodal values as:

$$
\begin{bmatrix} u_2 \\ u_3 \\ u_4 \end{bmatrix}
= -\frac{F_0 L^3}{E I_Y}
\begin{bmatrix} \frac{1}{256} \\ \frac{1}{128} \\ \frac{1}{256} \end{bmatrix} ,
\qquad (6.155)
$$

and the relative error in the middle of the beam is obtained as [1]:

$$
\text{relative error} = \left| \frac{\frac{1}{128} - \frac{1}{192}}{\frac{1}{192}} \right| \times 100 = 50.0\% .
\qquad (6.156)
$$

Writing the finite difference statement at the inner nodes $i = 2, \ldots, 8$ gives:

node 2: $\dfrac{E I_Y}{\Delta X^3} (u_4 - 4u_3 + 6u_2 - 4u_1 + u_0) = -q_0 \Delta X$, $\qquad$ (6.157)

node 3: $\dfrac{E I_Y}{\Delta X^3} (u_5 - 4u_4 + 6u_3 - 4u_2 + u_1) = -q_0 \Delta X$, $\qquad$ (6.158)

node 4: $\dfrac{E I_Y}{\Delta X^3} (u_6 - 4u_5 + 6u_4 - 4u_3 + u_2) = -q_0 \Delta X$, $\qquad$ (6.159)

node 5: $\dfrac{E I_Y}{\Delta X^3} (u_7 - 4u_6 + 6u_5 - 4u_4 + u_3) = -q_0 \Delta X$, $\qquad$ (6.160)

node 6: $\dfrac{E I_Y}{\Delta X^3} (u_8 - 4u_7 + 6u_6 - 4u_5 + u_4) = -q_0 \Delta X$, $\qquad$ (6.161) node 7: $\dfrac{EI_Y}{\Delta X^3}(u_9 - 4u_8 + 6u_7 - 4u_6 + u_5) = -q_0\Delta X$,   (6.162)

node 8: $\dfrac{EI_Y}{\Delta X^3}(u_{10} - 4u_9 + 6u_8 - 4u_7 + u_6) = -q_0\Delta X$,   (6.163)

or under consideration of the boundary conditions, i.e. $u_1 = u_9 = 0$, $u_0 = u_2$ and $u_{10} = u_8$, in matrix notation:

$$
\begin{bmatrix}
7 & -4 & 1 & 0 & 0 & 0 & 0 \\
-4 & 6 & -4 & 1 & 0 & 0 & 0 \\
1 & -4 & 6 & -4 & 1 & 0 & 0 \\
0 & 1 & -4 & 6 & -4 & 1 & 0 \\
0 & 0 & 1 & -4 & 6 & -4 & 1 \\
0 & 0 & 0 & 1 & -4 & 6 & -4 \\
0 & 0 & 0 & 0 & 1 & -4 & 7
\end{bmatrix}
\begin{bmatrix}
u_2 \\ u_3 \\ u_4 \\ u_5 \\ u_6 \\ u_7 \\ u_8
\end{bmatrix}
= -\frac{F_0\Delta X^3}{EI_Y}
\begin{bmatrix}
0 \\ 0 \\ 0 \\ 1 \\ 0 \\ 0 \\ 0
\end{bmatrix}.
\tag{6.164}
$$

The solution of this linear system of equations gives the unknown nodal values as:

$$
\begin{bmatrix}
u_2 \\ u_3 \\ u_4 \\ u_5 \\ u_6 \\ u_7 \\ u_8
\end{bmatrix}
= -\frac{F_0 L^3}{EI_Y}
\begin{bmatrix}
\frac{1}{1024} \\
\frac{3}{1024} \\
\frac{5}{1024} \\
\frac{3}{512} \\
\frac{5}{1024} \\
\frac{3}{1024} \\
\frac{1}{1024}
\end{bmatrix},
\tag{6.165}
$$

and the relative error in the middle of the beam is obtained as [1]:

$$
\text{relative error} = \left| \frac{\frac{3}{512} - \frac{1}{192}}{\frac{1}{192}} \right| \times 100 = 12.5\% .
\tag{6.166}
$$

3.22 Finite difference approximation of a cantilevered beam based on five domain nodes with imposed tip displacement

The finite difference discretization of the cantilevered beam is shown in Fig. 6.10 for five domain nodes.

Evaluation of the finite difference approximation of the fourth-order differential equation according to Eq. (3.9) at the inner nodes $i = 2, \ldots, 4$ gives:

node 2: $\dfrac{EI_Y}{\Delta X^3}(u_4 - 4u_3 + 6u_2 - 4u_1 + u_0) = 0$,   (6.167)

node 3: $\dfrac{EI_Y}{\Delta X^3}(u_5 - 4u_4 + 6u_3 - 4u_2 + u_1) = 0$,   (6.168)

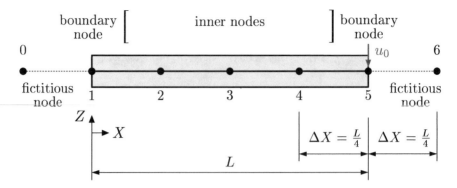

Fig. 6.10 Finite difference discretization of the cantilevered Euler–Bernoulli beam (imposed displacement) based on five grid nodes

$$\text{node 4:} \quad \frac{EI_Y}{\Delta X^3}\left(u_6 - 4u_5 + 6u_4 - 4u_3 + u_2\right) = 0. \tag{6.169}$$

Under consideration of the boundary conditions at the left-hand end, i.e., $u_1 = 0$ and $u_0 = u_2$, and the conditions at the right-hand boundary, i.e., $u_5 = -u_0$ and $u_6 = 2u_5 - u_4$, the following matrix scheme can be stated:

$$\begin{bmatrix} 7 & -4 & 1 \\ -4 & 6 & -4 \\ 1 & -4 & 5 \end{bmatrix} \begin{bmatrix} u_2 \\ u_3 \\ u_4 \end{bmatrix} = -u_0 \begin{bmatrix} 0 \\ 1 \\ -2 \end{bmatrix}. \tag{6.170}$$

The solution of this linear system of equations gives the unknown nodal values as:

$$\begin{bmatrix} u_2 \\ u_3 \\ u_4 \end{bmatrix} = -u_0 \begin{bmatrix} \frac{1}{11} \\ \frac{7}{22} \\ \frac{7}{11} \end{bmatrix} = -u_0 \begin{bmatrix} 0.09090909 \\ 0.31818182 \\ 0.63636364 \end{bmatrix}. \tag{6.171}$$

The analytical solution can be taken from [2] as

$$u(X) = \left[\frac{1}{2}\left(\frac{X}{L}\right)^3 - \frac{3}{2}\left(\frac{X}{L}\right)^2\right] u_0, \tag{6.172}$$

and the relative error is obtained, for example, at $X = \frac{3L}{4}$ as:

$$\text{relative error} = \left|\frac{\frac{7}{11} - \frac{81}{128}}{\frac{81}{128}}\right| \times 100 = 0.56\%. \tag{6.173}$$

6.4 Answers for Problems from Chap. 4

4.2 Finite difference approximation of a beam fixed at both ends
(a) Single force case

$$u_Z(0.5L) = \frac{\left(32EI_Y + k_s AGL^2\right) LF_0}{2k_s AG \left(k_s AGL^2 + 64EI_Y\right)} \quad \text{(5 nodes)}, \qquad (6.174)$$

$$u_Z(0.5L) = \frac{\left(512EI_Y + 11k_s AGL^2\right) LF_0}{2k_s AG \left(k_s AGL^2 + 1024EI_Y\right)} \quad \text{(17 nodes)}. \qquad (6.175)$$

Relative error is -95.385% for 5 nodes and -65.104% for 17 nodes.

(b) Distributed load case

$$u_Z(0.5L) = \frac{\left(32EI_Y + k_s AGL^2\right) L^2 q_0}{4k_s AG \left(k_s AGL^2 + 64EI_Y\right)} \quad \text{(5 nodes)}, \qquad (6.176)$$

$$u_Z(0.5L) = \frac{\left(512EI_Y + 11k_s AGL^2\right) L^2 q_0}{4k_s AG \left(k_s AGL^2 + 1024EI_Y\right)} \quad \text{(17 nodes)}. \qquad (6.177)$$

Relative error is -95.274% for 5 nodes and -64.267% for 17 nodes.

4.3 Convergence of finite difference approximation of a simply supported Timoshenko beam
Only the inner nodes are used to derive the finite difference approximation. The following general solutions consider that the load is opposed to the positive Z-direction. To calculate the numerical values, it is only required to insert the absolute value of q.

$$u_Z(0.5L) = -\frac{\left(32EI_Y + 3k_s AGL^2\right) L^2 q_0}{4k_s AG \left(k_s AGL^2 + 64EI_Y\right)} \quad \text{(5 nodes)}, \qquad (6.178)$$

$$u_Z(0.5L) = -\frac{\left(512EI_Y + 53k_s AGL^2\right) L^2 q_0}{4k_s AG \left(k_s AGL^2 + 1024EI_Y\right)} \quad \text{(13 nodes)}, \qquad (6.179)$$

$$u_Z(0.5L) = -\frac{\left(1936EI_Y + 201k_s AGL^2\right) L^2 q_0}{8k_s AG \left(k_s AGL^2 + 1936EI_Y\right)} \quad \text{(23 nodes)}, \qquad (6.180)$$

$$u_Z(0.5L) = -\frac{\left(2048EI_Y + 213k_s AGL^2\right) L^2 q_0}{4k_s AG \left(k_s AGL^2 + 4096EI_Y\right)} \quad \text{(33 nodes)}, \qquad (6.181)$$

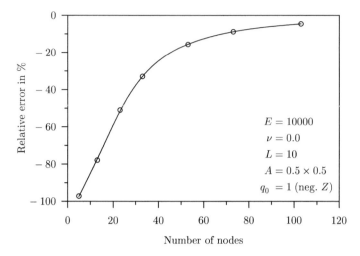

Fig. 6.11 Convergence rate of the finite difference approximation for a simply supported Timoshenko beam with distributed load

$$u_Z(0.5L) = -\frac{\left(5408EI_Y + 563k_sAGL^2\right)L^2 q_0}{4k_sAG\left(k_sAGL^2 + 10816EI_Y\right)} \quad (53 \text{ nodes}), \qquad (6.182)$$

$$u_Z(0.5L) = -\frac{\left(31104EI_Y + 3239k_sAGL^2\right)L^2 q_0}{12k_sAG\left(k_sAGL^2 + 20736EI_Y\right)} \quad (73 \text{ nodes}), \qquad (6.183)$$

$$u_Z(0.5L) = -\frac{\left(124848EI_Y + 13003k_sAGL^2\right)L^2 q_0}{24k_sAG\left(k_sAGL^2 + 41616EI_Y\right)} \quad (103 \text{ nodes}). \qquad (6.184)$$

The convergence rate is shown in Fig. 6.11.

4.4   Convergence of finite difference approximation of a cantilevered Timoshenko beam

(a) Solutions for dim. $= 2 \times$ (domain nodes -2). The following general solutions consider that the load is opposed to the positive Z-direction.

$$u_Z(L) = -\frac{\left(4096E^2I_Y^2 + 1248EI_Yk_sAGL^2 + 25k_s^2A^2G^2L^4\right)L^2 q_0}{4k_sAG\left(32EI_Y + k_sAGL^2\right)\left(k_sAGL^2 + 64EI_Y\right)} \quad (5 \text{ nodes}),$$
$$\qquad (6.185)$$

$$u_Z(L) = -\frac{\left(32EI_Y + 11k_sAGL^2\right)LF_0}{k_sAG\left(32EI_Y + k_sAGL^2\right)} \quad (5 \text{ nodes}). \qquad (6.186)$$

The convergence rate is shown in Fig. 6.12.

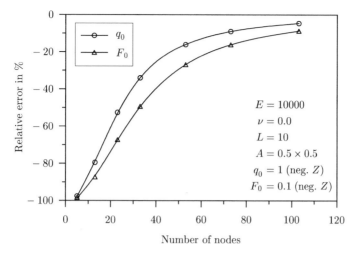

Fig. 6.12 Convergence rate of the finite difference approximation for a cantilevered Timoshenko beam with distributed or point load; dim. = $2 \times (\text{domain nodes} - 2)$

$$
\begin{bmatrix}
0 & -\frac{2EI_y}{\Delta X} - k_s AG \Delta X & -\frac{k_s AG}{2} & \frac{EI_y}{\Delta X} & 0 & 0 \\[2mm]
-\frac{2k_s AG}{\Delta X} & 0 & \frac{k_s AG}{\Delta X} & +\frac{k_s AG}{2} & 0 & 0 \\[2mm]
\frac{k_s AG}{2} & \frac{EI_y}{\Delta X} & 0 & -\frac{2EI_y}{\Delta X} - k_s AG \Delta X & -\frac{k_s AG}{2} & \frac{EI_y}{\Delta X} \\[2mm]
\frac{k_s AG}{\Delta X} & -\frac{k_s AG}{2} & -\frac{2k_s AG}{\Delta X} & 0 & \frac{k_s AG}{\Delta X} & \frac{k_s AG}{2} \\[2mm]
0 & 0 & \frac{2k_s AG}{3} & \frac{2EI_y}{3\Delta X} - \frac{k_s AG \Delta X}{9} & -\frac{2k_s AG}{3} & -\frac{2EI_y}{3\Delta X} - \frac{5k_s AG \Delta X}{9} \\[2mm]
0 & 0 & \frac{2k_s AG}{3\Delta X} & -\frac{4k_s AG}{9} & -\frac{2k_s AG}{3\Delta X} & +\frac{2k_s AG}{9}
\end{bmatrix}
\begin{bmatrix}
u_2 \\[2mm] \phi_2 \\[2mm] u_3 \\[2mm] \phi_3 \\[2mm] u_4 \\[2mm] \phi_4
\end{bmatrix}
=
\begin{bmatrix}
0 \\[2mm] q_0 \Delta X \\[2mm] 0 \\[2mm] q_0 \Delta X \\[2mm] 0 \\[2mm] q_0 \Delta X
\end{bmatrix}
\tag{6.187}
$$

In the case of the single force, the right-hand side of Eq. (6.187) must be replaced by the following column matrix:

$$\left[0\ 0\ 0\ 0\ \frac{F_0 \Delta X}{3}\ \frac{2F}{3}\right]^{\mathrm{T}} . \tag{6.188}$$

The value of the deflection at the right-hand end can be obtained from the solution of the linear system of equations as:

$$u(X = L) = -\frac{u_3}{3} + \frac{4u_4}{3} - \frac{2\Delta X \phi_3}{9} + \frac{8\Delta X \phi_4}{9} + \frac{2\Delta X Q_5}{3k_s AG} . \tag{6.189}$$

(b) Solutions for dim. $= 2 \times$ (domain nodes $- 1$).
The fictitious node at the right-hand boundary can be eliminated for example in the case of five domain nodes by the kinematic relationship

$$\phi_5 = -\frac{\mathrm{d}u_Z(L)}{\mathrm{d}X} + \gamma_{XZ}(L) \approx -\frac{u_6 - u_4}{2\Delta X} + \frac{\tau_{XZ}(L)}{G} = -\frac{u_6 - u_4}{2\Delta X} + \frac{Q_Z(L)}{k_s AG} , \tag{6.190}$$

or

$$u_6 = u_4 - \phi_5 2\Delta X + \frac{Q_Z(L)}{k_s AG} . \tag{6.191}$$

The condition for the bending moment at the free boundary gives: $\phi_6 = \phi_4$.
In the case of this approach, the system of equations given in Eq. (6.187) is extended by two columns and rows whereas the last two rows reads as:

$$\begin{bmatrix} \cdots & 0 & \frac{EI_Y}{\Delta X} & 0 & -\frac{2EI_Y}{\Delta X} \\ \cdots & \frac{2k_s AG}{\Delta X} & 0 & -\frac{2k_s AG}{\Delta X} & -2k_s AG \end{bmatrix} \begin{bmatrix} u_5 \\ \phi_5 \end{bmatrix} = \begin{bmatrix} 0 \\ \frac{q_0 \Delta X}{2} \end{bmatrix} , \tag{6.192}$$

or in the case of the single force as:

$$\begin{bmatrix} \cdots & 0 & \frac{EI_Y}{\Delta X} & 0 & -\frac{2EI_Y}{\Delta X} \\ \cdots & \frac{2k_s AG}{\Delta X} & 0 & -\frac{2k_s AG}{\Delta X} & -2k_s AG \end{bmatrix} \begin{bmatrix} u_5 \\ \phi_5 \end{bmatrix} = \begin{bmatrix} -Q_Z(L)\Delta X \\ F_0 - 2Q_Z(L) \end{bmatrix} . \tag{6.193}$$

The convergence rate is shown in Fig. 6.13.
    Let us finally remind that the equilibrium between the internal reaction (shear force) and external load at the right boundary gives: $Q_Z(L) = -F_0$.

$$u_Z(L) = -\frac{\left(224EI_Y + 55k_s AGL^2\right) L^2 q_0}{8k_s AG \left(64EI_Y + k_s AGL^2\right)} \quad \text{(5 nodes)}, \tag{6.194}$$

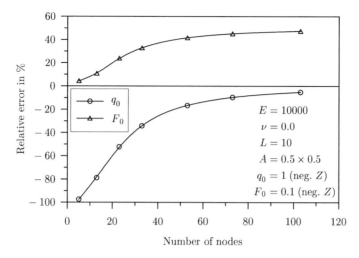

Fig. 6.13 Convergence rate of the finite difference approximation for a cantilevered Timoshenko beam with distributed or point load; dim. $= 2 \times$ (domain nodes $- 1$)

$$u_Z(L) = -\frac{\left(3072E^2I_Y^2 + 1024EI_Yk_sAGL^2 + 11k_s^2A^2G^2L^4\right)LF_0}{32k_sAGEI_Y\left(64EI_Y + k_sAGL^2\right)} \quad \text{(5 nodes)}.$$

$$\text{(6.195)}$$

6.5 Answers for Problems from Chap. 5

5.1 Comparison between analytical and layer-wise integration of the bending stiffness

The relative error as a function of the layer number for different size of the plastic zone is shown in Fig. 6.14.

5.2 Investigation of the proportions of the relative bending stiffness

The absolute and relative difference between the proportions of the relative bending stiffness in the elastic range within the layered approach is shown in Fig. 6.15.

5.3 Elasto-plastic finite difference solution: influence of layer number

The numerical results for the deformed shape and a comparison with the analytical solution is presented in Table 6.3.

5.4 Elasto-plastic finite difference solution: influence of load increment

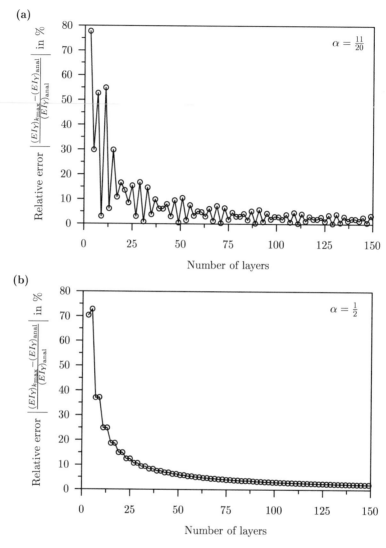

Fig. 6.14 Relative error as a function of the layer number for different size of the plastic zone

The numerical results for the deformed shape and a comparison with the analytical solution is shown in Table 6.4.

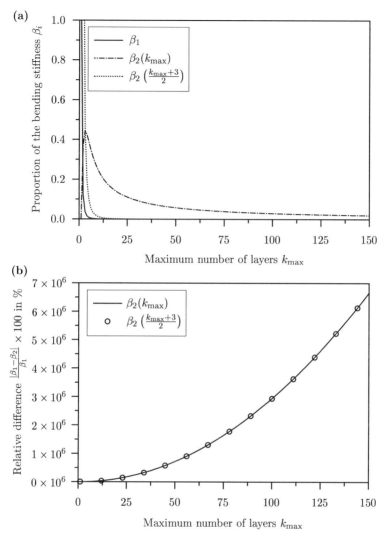

Fig. 6.15 a Absolute difference and **b** relative difference between the proportions of the relative bending stiffness in the elastic range within the layered approach

Table 6.3 Numerical results for deformed shape and comparison with analytical solution: influence of k_{max}

$k_{max} = 11$		$k_{max} = 101$	$k_{max} = 1001$	$k_{max} = 10001$
Displacement $u_Z(X)$				
$\frac{X}{L} = 0.0$	0.0	0.0	0.0	0.0
$\frac{X}{L} = 0.25$	-0.110762	-0.120338	-0.121357	-0.121390
$\frac{X}{L} = 0.5$	-0.147682	-0.160451	-0.161810	-0.161853
Relative Error in %				
$\frac{X}{L} = 0.0$	–	–	–	–
$\frac{X}{L} = 0.25$	2.933360	5.459040	6.352147	6.380597
$\frac{X}{L} = 0.5$	2.933148	5.459271	6.352380	6.380830

Table 6.4 Numerical results for deformed shape and comparison with analytical solution: influence of load increment

$\Delta M = 1/5$		$\Delta M = 1/8$	$\Delta M = 1/10$
Displacement $u_Z(X)$			
$\frac{X}{L} = 0.0$	0.0	0.0	0.0
$\frac{X}{L} = 0.25$	-0.128549	-0.119502	-0.120338
$\frac{X}{L} = 0.5$	-0.171399	-0.159336	-0.160451
Relative Error in %			
$\frac{X}{L} = 0.0$	–	–	–
$\frac{X}{L} = 0.25$	12.654530	4.726031	5.459040
$\frac{X}{L} = 0.5$	12.654777	4.726261	5.459271

References

1. Öchsner A (2014) Elasto-plasticity of frame structure elements: modeling and simulation of rods and beams. Springer, Berlin
2. Öchsner A (2020) Computational statics and dynamics: an introduction based on the finite element method. Springer, Singapore

Appendix A
Mathematics

A.1 Greek Alphabet

See Table A.1

A.2 Frequently Used Constants

$$\pi = 3.14159\,,$$
$$e = 2.71828\,,$$
$$\sqrt{2} = 1.41421\,,$$
$$\sqrt{3} = 1.73205\,,$$
$$\sqrt{5} = 2.23606\,,$$
$$\sqrt{e} = 1.64872\,,$$
$$\sqrt{\pi} = 1.77245\,.$$

© The Author(s), under exclusive license to Springer Nature Switzerland AG 2021   151
A. Öchsner, *Structural Mechanics with a Pen*,
https://doi.org/10.1007/978-3-030-65892-2

Table A.1 The Greek alphabet

Name	Small letters	Capital letters
Alpha	α	A
Beta	β	B
Gamma	γ	Γ
Delta	δ	Δ
Epsilon	ϵ	E
Zeta	ζ	Z
Eta	η	H
Theta	θ, ϑ	Θ
Iota	ι	I
Kappa	κ	K
Lambda	λ	Λ
Mu	μ	M
Ni	ν	N
Xi	ξ	Ξ
Omicron	o	O
Pi	π	Π
Rho	ρ, ϱ	P
Sigma	σ	Σ
Tau	τ	T
Upsilon	υ	Υ
Phi	ϕ, φ	Φ
Chi	χ	X
Psi	ψ	Ψ
Omega	ω	Ω

A.3   Special Products

$$(X + Y)^2 = X^2 + 2XY + Y^2 , \tag{A.1}$$

$$(X - Y)^2 = X^2 - 2XY + Y^2 , \tag{A.2}$$

$$(X + Y)^3 = X^3 + 3X^2Y + 3XY^2 + Y^3 , \tag{A.3}$$

$$(X - Y)^3 = X^3 - 3X^2Y + 3XY^2 - Y^3 , \tag{A.4}$$

$$(X + Y)^4 = X^4 + 4X^3Y + 6X^2Y^2 + 4XY^3 + Y^4 , \tag{A.5}$$

$$(X - Y)^4 = X^4 - 4X^3Y + 6X^2Y^2 - 4XY^3 + Y^4 . \tag{A.6}$$

A.4 Derivatives

- $\dfrac{d}{dX}\left(\dfrac{1}{X}\right) = -\dfrac{1}{X^2}$

- $\dfrac{d}{dX} X^n = n \times X^{n-1}$

- $\dfrac{d}{dX}\sqrt[n]{X} = \dfrac{1}{n \times \sqrt[n]{X^{n-1}}}$

- $\dfrac{d}{dX}\sin(X) = \cos(X)$

- $\dfrac{d}{dX}\cos(X) = -\sin(X)$

- $\dfrac{d}{dX}\ln(X) = \dfrac{1}{X}$

- $\dfrac{d}{dX}|X| = \begin{cases} -1 \text{ for } X < 0 \\ 1 \text{ for } X > 0 \end{cases}$

- $\dfrac{d}{dX}(f(X) \times g(X)) = \dfrac{df(X)}{dX}g(X) + f(X)\dfrac{dg(X)}{dX}$ (product rule)

- $\dfrac{d}{dX}\left(\dfrac{f(X)}{g(X)}\right) = \dfrac{df(X)/dX \times g(X) - f(X) \times dg(X)/dX}{[g(X)]^2}$ (quotient rule)

A.5 Integrals

The indefinite integral or antiderivative $F(X) = \int f(X)dX + c$ of a function $f(X)$ is a differentiable function $F(X)$ whose derivative is equal to $f(X)$, i.e., $\frac{dF(X)}{dX} = f(X)$. The definite integral of a continuous real-valued function $f(X)$ on a closed interval $[a, b]$, i.e., $\int_a^b f(d)dX = F(b) - F(a)$, is represented by the area under the curve $f(X)$ from $X = a$ to $X = b$.

Some selected antiderivatives (c: arbitrary constant of integration):

- $\int e^X dX = e^X + c$
- $\int \sqrt{X}dX = \frac{2}{3} X^{\frac{3}{2}} + c$
- $\int \sin(X)dX = -\cos(X) + c$
- $\int \cos(X)dX = \sin(X) + c$
- $\int \sin(\alpha X) \cdot \cos(\alpha X)dX = \dfrac{1}{2\alpha}\sin^2(\alpha X) + c$
- $\int \sin^2(\alpha X)dX = \dfrac{1}{2}(X - \sin(\alpha X)\cos(\alpha X)) + c = \dfrac{1}{2}(X - \frac{1}{2\alpha}\sin(2\alpha X)) + c$
- $\int \cos^2(\alpha X)dX = \dfrac{1}{2}(X + \sin(\alpha X)\cos(\alpha X)) + c = \dfrac{1}{2}(X + \frac{1}{2\alpha}\sin(2\alpha X)) + c$

Fig. A.1 Graphical
representation of the definite
integral as the area A under
the graph of $f(X)$

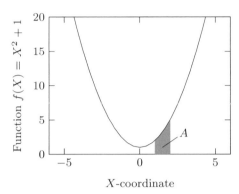

A.1 Example: Indefinite and definite integral

Calculate the indefinite and definite integral of $f(X) = X^2 + 1$. The definite integral
is to be calculated in the interval $[1, 2]$. Furthermore, give a graphical interpretation
of the definite integral.

A.1 Solution

Indefinite Integral:

$$F(X) = \int (X^2 + 1)\mathrm{d}X = \tfrac{X^3}{3} + X + c.$$ (A.7)

Definite Integral:

$$\int_1^2 (X^2 + 1)\mathrm{d}X = \left[\tfrac{X^3}{3} + X\right]_1^2 = \tfrac{10}{3}.$$ (A.8)

The graphical interpretation of the definite integral is shown in Fig. A.1.

Appendix B
Mechanics

B.1 Centroids

The coordinates (Z_S, Y_S) of the centroid S of the plane surface shown in Fig. B.1 can be expressed as

$$Z_S = \frac{\int Z \, dA}{\int dA}, \tag{B.1}$$

$$Y_S = \frac{\int Y \, dA}{\int dA}, \tag{B.2}$$

where the integrals $\int Z \, dA$ and $\int Y \, dA$ are known as the first moments of area[1]. In the case of surfaces composed of n simple shapes, the integrals can be replaced by summations to obtain:

$$Z_S = \frac{\sum\limits_{i=1}^{n} Z_i A_i}{\sum\limits_{i=1}^{n} A_i}, \tag{B.3}$$

$$Y_S = \frac{\sum\limits_{i=1}^{n} Y_i A_i}{\sum\limits_{i=1}^{n} A_i}. \tag{B.4}$$

---

[1] A better expression would be moment of surface since area means strictly speaking the measure of the size of the surface which is different to the surface itself.

© The Author(s), under exclusive license to Springer Nature Switzerland AG 2021
A. Öchsner, *Structural Mechanics with a Pen*,
https://doi.org/10.1007/978-3-030-65892-2

Fig. B.1 Plane surface with
centroid S

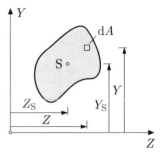

B.2   Second Moment of Area

The second moment of area[2] or the second area moment is a geometrical property
of a surface which reflects how its area elements are distributed with regard to an
arbitrary axis. The second moments of area for an arbitrary surface with respect to
an arbitrary Cartesian coordinate system (see Fig. B.1) are generally defined as:

$$I_Y = \int_A Z^2 \mathrm{d}A \,, \tag{B.5}$$

$$I_Z = \int_A Y^2 \mathrm{d}A \,. \tag{B.6}$$

These quantities are normally used in the context of plane bending of symmetrical
cross sections. For unsymmetrical bending, the product moment of area is addition-
ally required:

$$I_{YZ} = - \int_A Y Z \mathrm{d}A \,. \tag{B.7}$$

B.3   Parallel-Axis Theorem

The parallel-axis theorem gives the relationship between the second moment of area
with respect to a centroidal axis (Z_1, Y_1) and the second moment of area with respect
to any parallel axis[3] (Z, Y). For the rectangular shown in Fig. B.2, the relations can
be expressed as:

---

[2]The second moment of area is also called in the literature the second moment of inertia. However,
the expression moment of inertia is in context of properties of surfaces misleading since no mass
or movement is involved.

[3]This arbitrary axis can be for example the axis trough the common centroid S of a composed
surface.

Fig. B.2 Configuration for
the parallel-axis theorem

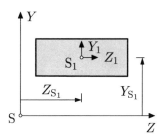

$$I_Y = I_{Y_1} + Z_{S_1}^2 \times A_1 \,, \tag{B.8}$$

$$I_Z = I_{Z_1} + Y_{S_1}^2 \times A_1 \,, \tag{B.9}$$

$$I_{ZY} = I_{Z_1 Y_1} - Z_{S_1} Y_{S_1} \times A_1 \,. \tag{B.10}$$

Index

© The Author(s), under exclusive license to Springer Nature Switzerland AG 2021   159
A. Öchsner, *Structural Mechanics with a Pen*,
https://doi.org/10.1007/978-3-030-65892-2

Printed in the United States
by Baker & Taylor Publisher Services